Contents

About your **BTEC Level 2 First Construction**

Credits

The publisher would like to thank the following individuals and organizations for their kind permission to reproduce photographs:

About the authors

Simon Topliss

Simon Topliss (BSc (Hons) P.G.C.E. M.I.F.L.) has worked for Edexcel for over six years as an External Verifier and Examiner. He has taught for over eleven years on a variety of Construction Technical Qualifications at Further and Higher Education levels. Simon has developed a suite of published resources for Levels 2 and 3, written unit specifications for the Firsts, Nationals and Higher Nationals, and is Senior Examiner and Principal Moderator on the CBE Diploma Level 2.

Mark Doyle

Mark Doyle has worked for Edexcel as a Principal Moderator, Senior Examiner and qualification writer for over ten years. He has taught in Further Education colleges since 1994 and has many years of college management experience. Mark is a co-author of the Edexcel Student Book and Tutor Resource Pack for the Level 2 Diploma in Construction and the Built Environment.

Ashley Stokes

Ashley Stokes has worked for more than 12 years as a construction tutor at York College, and has recently joined the Edexcel team as a Moderator on the Diploma in Construction and the Built Environment.

About your BTEC Level 2 First Construction

Choosing to study for a BTEC Level 2 First Construction qualification is a great decision to make for lots of reasons.

The construction industry is an expanding industry in the UK and many companies within it are looking to recruit qualified workers who are eager to continue their learning. This means that the demand for well qualified people to work within the construction industry is high. With a range of high-profile and local construction projects over the whole country, there will be even more opportunities for people with construction related qualifications.

Your BTEC Level 2 First Construction is a **vocational** or **work-related** qualification. This doesn't mean that it will give you *all* the skills you need to do a job, but it does mean that you'll have the opportunity to gain specific knowledge, understanding and skills that are relevant to your chosen subject or area of work.

What will you be doing?

The qualification is structured into **mandatory units** (ones you must do, marked as 'M') and **optional units** (ones you can choose to do, marked as 'O'). This book contains all 20 units, so you can be sure that you are covered whichever qualification you are working towards.

* BTEC Level 2 First Certificate in Construction: 2 mandatory units and an optional unit that provide a combined total of 15 credits

* BTEC Level 2 First Extended Certificate in Construction: 2 mandatory units and optional units that provide a combined total of 30 credits

* BTEC Level 2 First Diploma in Construction: 3 mandatory units and optional units that provide a combined total of 60 credits

Unit number	Credit value	Unit name	Cert	Ex. Cert	Diploma
1	5	Structure of the construction industry	M	M	M
2	5	Exploring health, safety and welfare in construction	M	M	M
3	5	Sustainability in the construction industry			M
8	5	Exploring carpentry and joinery	O	O	O
9	5	Performing joinery operations			O
10	5	Performing carpentry operations			O
11	5	Exploring trowel operations	O	O	O
12	5	Performing blockwork operations			O
13	5	Performing brickwork operations			O
14	5	Exploring painting and decorating	O	O	O
15	5	Performing paperhanging operations			O
16	5	Performing decorating operations			O

How to use this book

This book is designed to help you through your BTEC Level 2 First Construction course. It is divided into 12 units to reflect the units in the specification. This book contains many features that will help you use your skills and knowledge in work-related situations and assist you in getting the most from your course.

Introduction

These introductions give you a snapshot of what to expect from each unit – and what you should be aiming for by the time you finish it!

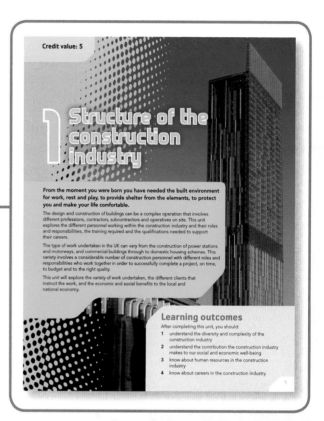

Assessment and grading criteria

This table explains what you must do in order to achieve each of the assessment criteria for each unit. For each assessment criterion, shown by the grade button **P1**, there is an assessment activity.

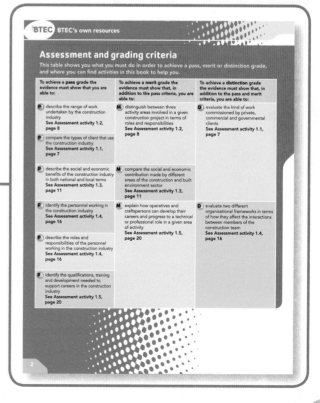

Assessment

Your tutor will set assignments throughout your course for you to complete. These may take the form of projects where you research, plan, prepare, and make a piece of practical work, case studies and presentations. The important thing is that you evidence your skills and knowledge to date.

Stuck for ideas? Daunted by your first assignment? These students have all been through it before…

How you will be assessed

This unit will be assessed by an internal assignment that will be designed and marked by the staff at your centre. Your assessment could be in the form of

- presentations
- case studies
- practical tasks
- written assignments

Ryan, 14 years old

I've been interested in construction for a while. Just by looking around me I already had a pretty good idea of the different types of building work that are done. This unit made me think about the areas that I had not realised were part of the construction industry, such as oil rigs and building and maintaining roads.

Before doing this unit I hadn't realised how many different jobs, roles and responsibilities were involved. We visited three different kinds of building site and met a range of people who work in the construction industry. I enjoyed learning about the different aspects of work that each role is responsible for and is involved in. It was a good area to explore because it made me think about the sorts of job that I might be interested in looking into when I leave education, and what qualifications I will need.

Over to you!

- Which areas of this unit might you find challenging?
- Which section of the unit are you most looking forward to?
- Which areas of this unit do you already have a basic understanding of?

Activities

There are different types of activities for you to do: assessment activities are suggestions for tasks that you might do as part of your assignment and will help you develop your knowledge, skills and understanding. Each of these has grading tips that clearly explain what you need to do in order to achieve a pass, merit or distinction grade.

There are also suggestions for activities that will give you a broader grasp of the industry, stretch your understanding and deepen your skills.

Assessment activity 1.3 P3 M2

The architect in your design office has been working with a client who has local and national offices which are to be refurbished with new shop fronts on a rolling programme.

1 Produce a report to describe the social and economic benefits of the refurbishment programme in both national and local terms. **P3**

2 Expand your report to consider the social and economic contribution made by different areas of the built environment sector, namely civil engineering, architectural design and housing development. **M2**

Grading tips

1 To achieve **P3** make sure that your local and national benefits are clearly two different areas that do not contain the same items.

2 To achieve **M2** make sure that you clearly distinguish between the sector areas. You might decide to display this information in a table.

Activity: Fire classification

Complete the right-hand column of Table 2.2 to show the UK and European classifications of fire. Note that no 'E' classification exists in the UK.

Table 2.2: Fire classification

Class	Type of fire
A	
B	
C	
D	
E – not a class in the UK	
F	

Personal, learning and thinking skills

Throughout your BTEC Level 2 First Construction course, there are lots of opportunities to develop your personal, learning and thinking skills. Look out for these as you progress.

Functional skills

It's important that you have good English, maths and ICT skills – you never know when you might need them, and employers will be looking for evidence that you have these skills.

Key terms

Technical words and phrases are easy to spot, and definitions are included. The terms and definitions are also in the glossary at the back of the book.

WorkSpace

Case studies provide snapshots of real workplace issues, and show how the skills and knowledge you develop during your course can help you in your career.

PLTS

Independent enquirer develop questions and ask them of the guest speaker

Reflective learner use what you learn on site to inform your report

Functional skills

English use your writing skills effectively to produce your assignment

Key term

Encapsulate – cover or enclose completely without disturbing.

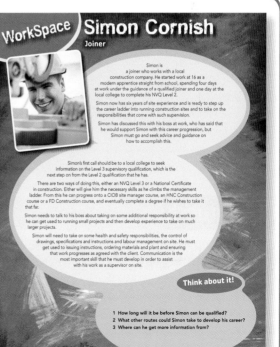

WorkSpace **Simon Cornish**
Joiner

Simon is a joiner who works with a local construction company. He started work at 16 as a modern apprentice straight from school, spending four days at work under the guidance of a qualified joiner and one day at the local college to complete his NVQ Level 2.

Simon now has six years of site experience and is ready to step up the career ladder into running construction sites and to take on the responsibilities that come with such supervision.

Simon has discussed this with his boss at work, who has said that he would support Simon with this career progression, but Simon must go and seek advice and guidance on how to accomplish this.

Simon's first call should be to a local college to seek information on the Level 3 supervisory qualification, which is the next step on from the Level 2 qualification that he has.

There are two ways of doing this, either an NVQ Level 3 or a National Certificate in construction. Either will give him the necessary skills as he climbs the management ladder. From this he can progress onto a CIOB site manager course, an HNC Construction course or a FD Construction course, and eventually complete a degree if he wishes to take it that far.

Simon needs to talk to his boss about taking on some additional responsibility at work so he can get used to running small projects and then develop experience to take on much larger projects.

Simon will need to take on some health and safety responsibilities, the control of drawings, specifications and instructions and labour management on site. He must get used to issuing instructions, ordering materials and plant and ensuring that work progresses as agreed with the client. Communication is the most important skill that he must develop in order to assist with his work as a supervisor on site.

Think about it!

1 How long will it be before Simon can be qualified?
2 What other routes could Simon take to develop his career?
3 Where can he get more information from?

21

Just checking

When you see this sort of activity, take stock! These quick activities and questions are there to check your knowledge. You can use them to see how much progress you've made.

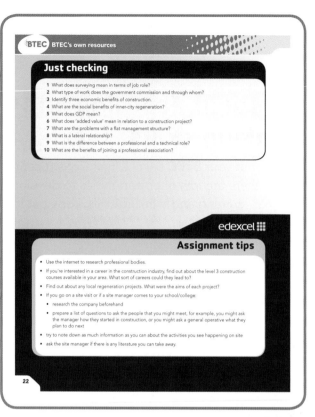

Edexcel's assignment tips

At the end of each chapter, you'll find hints and tips to help you get the best mark you can, such as the best websites to go to, checklists to help you remember processes and really useful facts and figures.

Have you read your BTEC Level 2 First Study Skills Guide? It's full of advice on study skills, putting your assignments together and making the most of being a BTEC Construction student.

Your book is just part of the exciting resources from Edexcel to help you succeed in your BTEC course. Visit **www.edexcel.com/BTEC** or **www.pearsonfe.co.uk/BTEC 2010** for more details.

1 Structure of the construction industry

From the moment you were born you have needed the built environment for work, rest and play, to provide shelter from the elements, to protect you and make your life comfortable.

The design and construction of buildings can be a complex operation that involves different professions, contractors, subcontractors and operatives on site. This unit explores the different personnel working within the construction industry and their roles and responsibilities, the training required and the qualifications needed to support their careers.

The type of work undertaken in the UK can vary from the construction of power stations and motorways, and commercial buildings through to domestic housing schemes. This variety involves a considerable number of construction personnel with different roles and responsibilities who work together in order to successfully complete a project, on time, to budget and to the right quality.

This unit will explore the variety of work undertaken, the different clients that instruct the work, and the economic and social benefits to the local and national economy.

Learning outcomes

After completing this unit, you should:

1 understand the diversity and complexity of the construction industry

2 understand the contribution the construction industry makes to our social and economic well-being

3 know about human resources in the construction industry

4 know about careers in the construction industry.

Assessment and grading criteria

This table shows you what you must do in order to achieve a pass, merit or distinction grade, and where you can find activities in this book to help you.

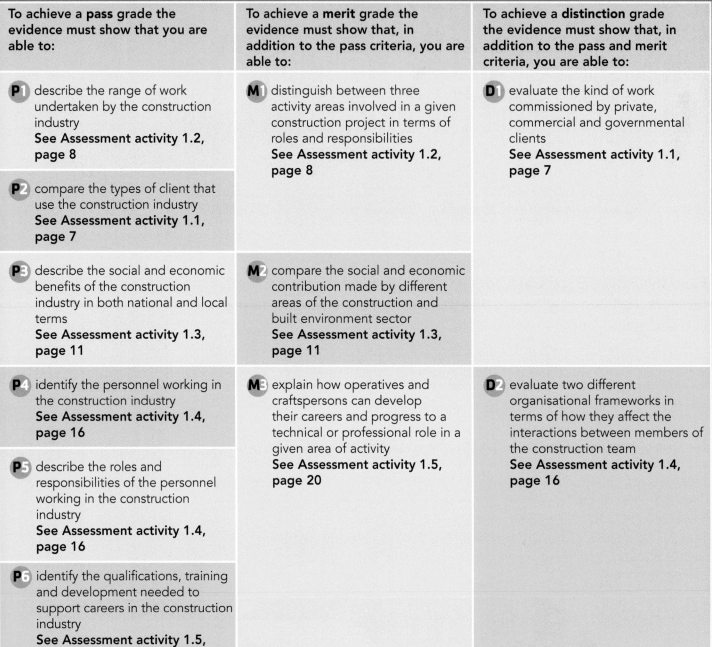

To achieve a **pass** grade the evidence must show that you are able to:	To achieve a **merit** grade the evidence must show that, in addition to the pass criteria, you are able to:	To achieve a **distinction** grade the evidence must show that, in addition to the pass and merit criteria, you are able to:
P1 describe the range of work undertaken by the construction industry **See Assessment activity 1.2, page 8**	**M1** distinguish between three activity areas involved in a given construction project in terms of roles and responsibilities **See Assessment activity 1.2, page 8**	**D1** evaluate the kind of work commissioned by private, commercial and governmental clients **See Assessment activity 1.1, page 7**
P2 compare the types of client that use the construction industry **See Assessment activity 1.1, page 7**		
P3 describe the social and economic benefits of the construction industry in both national and local terms **See Assessment activity 1.3, page 11**	**M2** compare the social and economic contribution made by different areas of the construction and built environment sector **See Assessment activity 1.3, page 11**	
P4 identify the personnel working in the construction industry **See Assessment activity 1.4, page 16**	**M3** explain how operatives and craftspersons can develop their careers and progress to a technical or professional role in a given area of activity **See Assessment activity 1.5, page 20**	**D2** evaluate two different organisational frameworks in terms of how they affect the interactions between members of the construction team **See Assessment activity 1.4, page 16**
P5 describe the roles and responsibilities of the personnel working in the construction industry **See Assessment activity 1.4, page 16**		
P6 identify the qualifications, training and development needed to support careers in the construction industry **See Assessment activity 1.5, page 20**		

How you will be assessed

This unit will be assessed by an internal assignment that will be designed and marked by the staff at your centre. Your assessment could be in the form of:

- presentations
- case studies
- practical tasks
- written assignments.

Ryan, 14 years old

I've been interested in construction for a while. Just by looking around me I already had a pretty good idea of the different types of building work that are done. This unit made me think about the areas that I had not realised were part of the construction industry, such as oil rigs and building and maintaining roads.

Before doing this unit I hadn't realised how many different jobs, roles and responsibilities were involved. We visited three different kinds of building site and met a range of people who work in the construction industry. I enjoyed learning about the different aspects of work that each role is responsible for and is involved in. It was a good area to explore because it made me think about the sorts of job that I might be interested in looking into when I leave education, and what qualifications I will need.

Over to you!

- Which areas of this unit might you find challenging?
- Which section of the unit are you most looking forward to?
- Which areas of this unit do you already have a basic understanding of?

1. Understand the diversity and complexity of the construction industry

Build up

One industry, many parts

The construction industry covers a great number of different activities. For example, a 3-D walk-through computer modelling of a building is a construction activity service provided by a design specialist. Take a few moments to reflect on what you know about the construction industry and create a list of all the different activities that you think form a part of it.

1.1 Activity areas

Building

Building is the general terminology used for construction; it covers a wide variety of construction work, for example, a garden wall, a single building or project, or a housing estate of 250 homes.

Architecture

This is the design side of construction. It relies on professional architects and technicians who provide a design service for clients and strive to produce a design that meets the needs of the client, while being eye-catching and sustainable. In the UK there is a wide range of architecture spanning hundreds of years, from old buildings that survived the Great Fire of London to modern cities, such as Milton Keynes.

Planning

Planning is the process of controlling a built environment project. Quality planning is essential to every construction or building project. A built environment project requires labour, plant, materials and sub-contractors. Planning is necessary to coordinate, control, forecast and communicate a contract programme. A plan enables smooth progress of production on site, highlights problems that need to be solved and helps keep everyone informed and updated.

Surveying

This is the activity of measuring the land, the building and any external works, as well as the setting out of the building, its associated external works and any other items needed to be placed in position relative to the architect's drawings. Surveying requires the use of tools such as levels, tapes and theodolites to measure lengths and angles, and to

Key term

Surveying – measuring an area to establish its size and shape.

calculate areas and volumes. The people involved in this activity are known as surveyors.

Civil and structural engineering

This activity area involves considerably larger projects that are not classified as buildings, for example:

- large earthworks such as motorway embankments and cuttings
- water works such as reservoirs and harbours
- large concrete works such as dams
- other infrastructure works such as roads and railways.

The people who are involved in this work are known as civil engineers.

Structural engineering is the process of using mathematics to design and detail a structure in order to make it stable, able to support its loads and safe for the occupants.

Building services engineering

This covers the services that support a building, for example, the heating, lighting and waste disposal. Services can be simple or very complex, for example, lift systems, escalators, intelligent boiler systems and automatic window-opening systems. Some services, such as fire alarms, continuously monitor the surroundings to keep occupants at the right temperature, comfortable and safe.

Facilities and estate management

A building needs to be maintained (looked after) during its life. Some items need to be replaced when they wear out or break, and others, such as fire extinguishers, must be replaced more frequently for safety reasons. An estate manager oversees a team of people who undertake the care and maintenance of a large commercial building, for example, a college or university.

Facilities management involves letting out to contractors various activities that occur within a building, and managing and monitoring their performance. For example, in a large hospital complex, the cleaning, portering, air conditioning and laundry may be run by several contractors under the direction and control of a facilities manager.

Highways engineers

Highways are the road networks of the UK. They include every size of road, for example, small estate roads, major trunk roads, dual carriageways and motorways. Highways engineers are responsible for the construction of this infrastructure in accordance with drawings and safety legislation. This involves setting out and monitoring the work, as well as the construction and maintenance of the roadway and any bridges over it.

Key term

Facilities – any activity required for the running and operation of a building.

Did you know?

There are approximately 250,000 miles of road in the UK.

1.2 Client types

Private

Private clients are sole traders or domestic clients who would like to have a house building altered, extended or maintained. They enter into private agreements with a builder to undertake the work. The private client may have an architect who has produced the drawings if planning or building regulations require this.

Commercial

A commercial client is a factory or business that needs to undertake building works in order to produce a product or process. For example, a fast food company needs an outlet to sell its products from. The outlet usually has to be built quickly, and to be maintenance-free and adaptable. Small industrial factory units are commercial buildings that provide products and services. These small-to-medium enterprises add considerably to the UK economy.

Public limited companies

A public limited company, such as a bank, is one that trades on the stock exchange and is owned by its shareholders. A bank could have hundreds of branches that all need to be maintained and looked after and upgraded by refurbishment from time to time. It is also likely to have a large headquarters in a major city, which will need similar work doing.

Did you know?

A public limited company has the same initials (PLC) as a private limited company.

The Government

The Government can issue work at three different levels: through local councils, devolved administrations (Welsh Assembly and Scottish Parliament), and central government. Local councils have duties to construct and maintain services. They can issue work such as the following:

- constructing schools
- maintaining highways
- replacing windows
- maintaining houses
- carrying out building works on council properties.

Devolved administrations can instruct major capital works such as the new parliament buildings or infrastructure works.

Central government departments,such as the Ministry of Defence (MoD), or bodies such as the Highways Agency or National Health Service, purchase a great deal of construction services, usually using intermediate companies who specialise in managing large building projects.

Assessment activity 1.1

You have just started your work experience week for a local architectural office. Your supervisor keeps mentioning the word 'client'. Your supervisor, seeing that you have a lack of knowledge in this area, asks you to complete the following tasks as part of your work experience report.

1 Identify and compare the different clients that use the construction industry to produce their construction projects. **P2**

2 Produce a short evaluation report of the kind of work that private, commercial and government clients commission. **D1**

Grading tips

1 To achieve **P2** you could include references to private, public and commercial organisations. Look around the built environment in your local area and try to identify the clients who would have commissioned the buildings so you can make a list.

2 To achieve **D1** you could give a detailed example of each of the three types of client, showing the differences between them and exploring why each one is typical of that type of client.

PLTS

Creative thinker generate ideas of who different clients could be

Independent enquirer explore the needs of different clients

Functional skills

ICT use a computer to research your evaluation report

1.3 The range of work undertaken

Table 1.1: Examples of the work undertaken for different types of client

Area of work	Description of work
Agricultural	Farm buildings to house livestock and store feed (designed by specialist building designers)
Commercial	Factory units, private hospitals, and production facilities for business enterprises
Educational	Buildings for schools, colleges, academies and universities. The Building Schools for the Future (BSF) programme is currently rebuilding a large number of schools throughout the UK.
Health	Private, National Health Service and Community Services hospitals are a large work activity area that continually changes with the development of new technologies.
Industrial	Large-scale heavy industrial factories, for example, oil refineries, provide a great number of different construction opportunities from civil engineering to construction of associated offices.
Public buildings	Local council public buildings, for example, council offices, town halls, libraries and distribution depots; national public buildings, for example art galleries, museums and other facilities
Recreational and leisure	Facilities for the community, for example, fitness clubs, leisure clubs, sports facilities, pavilions and community sports projects. They are sustainable developments that have been funded by government grants and the lottery.
Residential	Domestic housing i.e., new homes and extensions and affordable rented accommodation for Housing Associations
Retail	Retail out-of-town units, inner-city shopping refurbishments and high-street shop developments
Transport infrastructure	Roads, railways, tram systems, underground trains, motorways and bridges
Utilities	Installing, maintaining and repairing the infrastructure of the UK's utilities (water, electricity and gas)

PLTS

Independent enquirer develop questions and ask them of the guest speaker

Reflective learner use what you learn on site to inform your report

Functional skills

English use your writing skills effectively to produce your assignment

BTEC Assessment activity 1.2 **P**1 **M**1

Your tutor has asked in a guest speaker who is a local construction contracts manager to talk to you about the range of work that is undertaken by the construction industry. As a result they have invited the class to visit a local construction project where you will meet the various personnel involved in its construction.

1 Write a short assignment describing the range of work that is undertaken by the construction industry. **P**1

2 Site manager, health and safety officer and architect are three of the many roles within the construction industry. Produce a short report that clearly distinguishes between the roles and responsibilities for supervision, safety and design on a project. **M**1

Grading tips

1 To achieve **P**1 you could do some research into UK construction statistics.

2 To achieve **M**1 make sure that you clearly distinguish between the roles and responsibilities for each of the three different people. You might decide to display this information in a table, e.g., job, role, responsibility.

2. Know the contribution the construction industry makes to our social and economic well-being

2.1 The construction economy

Economic benefits of construction

The UK construction industry makes a valuable contribution to the UK economy. Constructing the built environment creates jobs before (in the planning stages) and during construction. Once the works have been completed, more people are able to move into an area. The construction of factories and retail units also provides business opportunities. The construction industry employs over 2 million people in many different roles and uses a great number of suppliers for materials and plant, along with many different specialist subcontractors. It is a massive economic operation.

What construction opportunities might be created by new energy technologies?

8

Inner-city regeneration

Many of our inner-city centres have been run down with little or no investment as large retail parks have been developed and constructed outside the city centre. This has led to a decline in the take-up of shops, office space and housing developments in city centres. With the injection of regional enterprise funding from Europe and the Government, many inner cities are experiencing a new lease of life. In Leeds, for example, a housing boom has led to the development of many multi-storey flats.

The housing market and property wealth

From the late 1990s to 2008, in the UK, the boom in housing brought about by the low cost of finance, coupled with the substantial rise in the average price of a house, produced wealth for many individuals and developers. The housing boom slowed considerably in 2009 as the global economic situation changed.

GDP

Gross domestic product (GDP) is a measure of the total expenditure of a country on goods and services within a certain time (normally a year). The construction industry contributes about 8 to 10 per cent of the UK's GDP.

Local and national contributions

Locally, the construction industry provides employment through projects, and creates a market for local plant and materials suppliers and all of the ancillary service areas that are required, such as waste skips and accommodation for workers.

Nationally, the construction industry undertakes work on infrastructure projects such as motorways and railway upgrades, sports stadia, for example, the Olympic Stadium, and contributes through the tax system to the wealth of the UK economy.

2.2 The social economy: social benefits of construction

Security

Users of the built environment need to feel safe in their homes, at work and during social hours. Building regulations and other legislation ensure that safety is designed into a structure.

A low crime rate will make an area more attractive to potential investors, who bring in wealth to develop an area.

Key term

Regeneration – redevelopment of land or the upgrading of older, rundown areas.

Did you know?

The UK construction industry:
- has an annual turnover of more than £100 billion
- accounts for almost 10% of the country's GDP
- employs two million people in more than 250,000 different companies.

The architect loved their design of this example of urban regeneration so much that they bought the top floor

Did you know?

All new buildings in the EU may have to comply with zero carbon building standards by 2019.

Added value

Adding value to a project can be done in several different ways, for example, by providing ample parking facilities that could be used when the offices are closed on Saturdays to support the local retail market and investment. Adding value is often a difficult item to consider when looking at the economic benefits of constructing the built environment.

Crime reduction

Investing in the built environment can help to reduce crime. Examples include investing in new infrastructure, community centres, CCTV systems, regular police patrols, designing out blind spots, the use of security lighting, and locked and gated communities.

Aesthetics

England has a rich history of excellent design. Historic buildings blend in with new skyscrapers, especially in the capital city of London. If you make an environment an attractive place to live then it will socially benefit the area as people then have a sense of belonging.

Urban renewal

Many areas have been subject to funding on 'enterprise action zones' and 'urban regeneration'. They use European Union (EU) and Government funds to demolish and rebuild old inner-city areas and breathe new life into them with a modern and attractive built environment.

Quality standards

Quality is achieved in the built environment through building regulations, planning legislation, British standards and commercial standards of quality design and construction.

Social contribution

The sustainability of a project can be enhanced by getting the local community involved and actively participating in the project. This gives a project a key sustainability factor. An example of this is the development of a sustainable housing estate where no cars are allowed.

Assessment activity 1.3 **BTEC** P3 M2

The architect in your design office has been working with a client who has local and national offices which are to be refurbished with new shop fronts on a rolling programme.

1 Produce a report to describe the social and economic benefits of the refurbishment programme in both national and local terms. **P3**

2 Expand your report to consider the social and economic contribution made by different areas of the built environment sector, namely civil engineering, architectural design and housing development. **M2**

Grading tips

1 To achieve **P3** make sure that your local and national benefits are clearly two different areas that do not contain the same items.

2 To achieve **M2** make sure that you clearly distinguish between the sector areas. You might decide to display this information in a table.

PLTS

Creative thinker get ideas from your own experience and those of friends

Independent enquirer plan your research and evaluate your findings

Functional skills

ICT use the internet to research the benefits of the programme

3. Know about human resources in the construction industry

3.1 Roles and responsibilities of members of the construction team

Figure 1.1 below shows some of the responsibilities of the client and their design team at the design stage of the work.

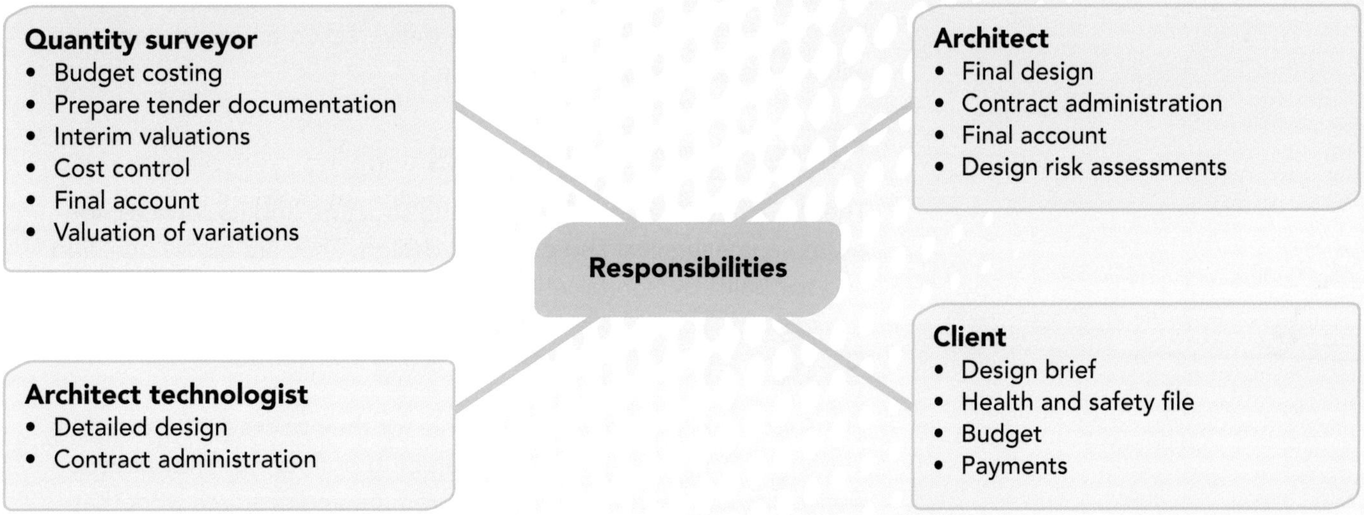

Quantity surveyor
- Budget costing
- Prepare tender documentation
- Interim valuations
- Cost control
- Final account
- Valuation of variations

Architect
- Final design
- Contract administration
- Final account
- Design risk assessments

Responsibilities

Architect technologist
- Detailed design
- Contract administration

Client
- Design brief
- Health and safety file
- Budget
- Payments

Fig. 1.1: Responsibilities of different team members in a built environment project

Building surveyor

A building surveyor may be responsible for a number of areas, for example, valuation of property for lenders, scheduling building defects, refurbishments and undertaking different types of building surveys, such as home information packs.

Land surveyor

This person deals with the measurement of land, which may involve the setting out of structures on the land in accordance with architects' drawings. Land surveyors often produce maps of the areas surveyed so that architects can make an accurate design for a structure.

Clerk of works

This person is often employed by a client to inspect, test and ensure that the construction work conforms to the quality standards that the designer requires. In effect they are the client's eyes and ears on the construction site.

Figure 1.2 on page 13 illustrates a typical construction site with the personnel who are involved with its supervision and running and their responsibilities during this period.

Estimator

This person is responsible for preparing the price or estimate of the cost of the work from the tender package prepared by the client's quantity surveyor. This estimate has to be carefully prepared to ensure that the best prices are obtained for materials and from subcontractors.

Buyer

As the name suggests, once a contract has been won, the buyer is responsible for purchasing all the materials from suppliers. The buyer has to ensure that the materials are delivered on time and within the estimator's costs stated in the tender.

Consulting engineers

These are engineers who work on the structure and its services, for example, mechanical and electrical design. They are highly qualified and specialist personnel who are required on complex projects.

Subcontractors

The subcontractor will be responsible for their package, for example, plastering, windows, electricals, plumbing or flooring. They have to work safely to the construction programme and produce work that meets the specification.

Contract manager's responsibilities
- Contract programme
- Sourcing resources
- Health and safety
- Site meetings
- Site inductions
- Coordination and control
- Administration

Safety officer's responsibilities
- Health and safety audits
- Inspections
- Monitoring risk assessments
- Health and safety testing

Site manager's responsibilities
- Labour, plant and materials organisation
- Records
- Safety
- Progress

Craftsperson's responsibilities
- Work safely and efficiently
- Produce quality workmanship
- Attend training

General operative's responsibilities
- Work safely
- Attend training

Fig. 1.2: Typical construction site set-up, detailing each role's responsibilities.

3.2 Interaction between team members

Simple organisational frameworks (top-down and flat structures)

Within an organisation, a management structure is used to control the business. This has various layers of management, with the most senior manager at the top of the tree.

Figure 1.3 below illustrates a top-down structure of management for a typical construction business. A top-down structure is more like a pyramid with several layers of management. Figure 1.4 below illustrates a flat structure of management. A flat management structure does not have as many layers as a top-down structure.

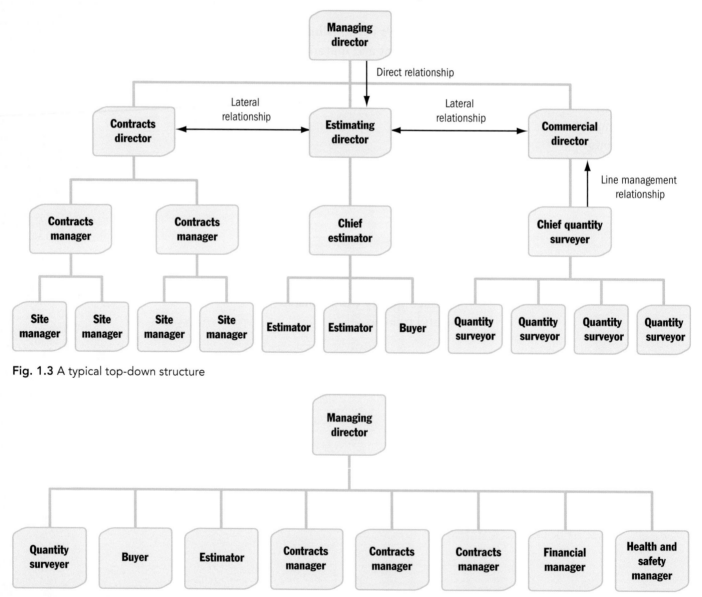

Fig. 1.3 A typical top-down structure

Fig. 1.4: A typical flat management structure

Direct and lateral relationships

The Managing Director in both of the structures has a direct relationship vertically downwards. This person is at the top of the structure and is the sole person in charge of the business. A lateral relationship is one which has people at the same level, for example, all the directors below the managing director are at the same level of responsibility. (See Fig. 1.3 on page 14.)

Service and line management relationships

Service or functional relationships involve people who work within a certain function, for example, the contracts director is responsible for the contracting side of the business, which includes the contracts managers and the site managers.

A line manager is the person to whom you report and take instructions from, for example, the quantity surveyor (QS) reports to the Chief QS who reports to the Commercial Director. (See Fig. 1.3 on page 14.)

Valuing others

Ensuring equality and diversity within an organisation helps to promote an excellent working environment for employees. Managers must value their workers and vice versa. Giving equal opportunities to all, and incorporating this into company policy, ensures that a company meets the employment laws of the UK.

Provision of safe working environment

Making workers safe is vital and is covered by the **HASAWA** (1974) which protects workers from harm. Employers have a legal duty to:

- keep all employees safe whilst at work

- provide all employees with personal protective equipment (PPE)

- undertake accident-prevention measures.

Key term

HASAWA – Health and Safety at Work etc. Act (1974).

Investment in staff training and development

Investing in staff and making them feel part of a team is a management skill that produces high levels of productivity from workers, and makes them feel motivated and competitive. Investors in People is a national award that recognises achievement in this area.

BTEC Assessment activity 1.4 P4 P5 D2

The marketing manager at the construction company you are working for has asked you to produce some materials that can be given to schools to promote recruitment for future apprentice positions.

1 Produce an A5 poster that identifies the people who work in the construction industry. **P4**

2 Expand the content into a leaflet covering each identified person to include a description of their roles and responsibilities. **P5**

3 Produce two visual charts of different organisational frameworks that show the interactions between construction team members on a typical construction site and head office. **D2**

Grading tips

If you get chance to visit a construction site, ask for a copy of the management structure.

1 To achieve **P4** you could differentiate between operatives, craftspeople, technical, supervisory, managerial and professional personnel.

2 To achieve **P5** you could make sure that you differentiate between all the different types of personnel.

3 To achieve **D2** you could include description of how each organisational framework operates and where its use is most appropriate.

4. Know about careers in the construction industry

4.1 Career paths

There are many career paths to follow within the construction industry. Some of them are discussed below.

Professional

This tends to be a designated role, for example, an architect, quantity surveyor or structural engineer who has qualified professionally by passing exams to become a member of a professional organisation. The professional organisations are discussed later in this section.

Technical

In management, this position is often known as a technician. It is an intermediate position and is someone who has a technical knowledge of construction but is not a fully qualified professional. For example, an architectural technologist produces CAD (computer-aided design) drawing details around the main design that has been created by the architect. This is a common career pathway within the construction industry.

Craftsperson

Apprenticeships, in which you learn the trade over several years, are a good way to become a fully qualified tradesperson in the traditional crafts of joinery and brickwork. Some of the more specialised modern apprenticeships include plumbing, electrical work, roofing and plastering. Craftspeople use their hands to produce work which is classified as 'skilled'.

General operative

This is essentially semi-skilled working and involves those construction operations that require manual labour. Excavations, concreting, drain laying and external works are all jobs that would be undertaken by general operatives. The skill level in general operatives is recognised if they can finish concrete or power float, or undertake duties above that of a general labourer.

Bridging arrangements for progression from craft to technical occupations

By undertaking a Level 3 qualification, people in craft roles have the potential to progress to a supervisory role, move onto a higher national certificate (HNC), foundation degree (FD) or degree, and become a fully qualified Contracts Manager, should they so wish. There is also a wide range of NVQs which provide the opportunity to learn whilst training at work to become qualified for a technical role.

4.2 Professional career pathways

There are several professional organisations that serve the UK construction industry:

- Royal Institute of British Architects (RIBA)
- Chartered Institute of Building Services Engineers (CIBSE)
- Institute of Civil Engineers (ICE)
- Royal Institute of Chartered Surveyors (RICS)
- Chartered Institute of Building (CIOB).

Key term

Apprenticeship – a programme of learning and qualifications, completed in the workplace and college or training centre, that gives young people the skills, knowledge and competence they need to progress in their chosen career or industry.

Key term

Professional – a person whose occupation requires specialist learning.

They provide professional status that is recognised worldwide. Table 1.2 illustrates the main professional bodies in the UK construction industry and lists the type of professional who could become a member of each organisation.

Table 1.2: The professional bodies for professionals in the construction industry

Professional body	Type of professional
Royal Institute of British Architects (RIBA)	Architect
Chartered Institute of Building Services Engineers (CIBSE)	Building services engineer
Institute of Civil Engineers (ICE)	Civil engineer
Royal Institute of Chartered Surveyors (RICS)	Building surveyor Land surveyor Quantity surveyor Site manager
Chartered Institute of Building (CIOB)	Contracts manager

4.3 Benefits of professional pathways

There are many benefits of professional career pathways. Some are outlined in Table 1.3 below.

Table 1.3: Benefits of professional career pathways

Benefit	What it means
Professional approach	Membership of a professional body helps to project the right image and approach to working with construction clients and shows that you follow the association's codes of conduct on how to behave.
Reputation	If you have professional letters after your name then clients are able to tell that you have spent time qualifying and meet professional standards.
Lifelong learning	Continuous professional development (CPD) has to be undertaken by members of associations so their knowledge of their industry remains up to date.
Advancement	Advancing through the levels of management and achieving a professional status is motivating and gives you a sense of achievement.
Promotion prospects	Professional roles are the highest point within a career progression and are a point to aim for.
Salary	A professional person is normally paid a salary higher than a person in a technical or craft position in line with the expertise, skills and years of training they have undertaken.
Position	Being a professional gives you a standing or status within an organisation.
Ability	The ability of a professional is supported by the organisation that they belong to, which has strict entrance qualifications and requires a certain level of experience to be attained before full membership is given.
Client relationships	Clients will use a professional because of the experience and knowledge that this brings to the design and construction of a project

4.4 Training and education

Training

There are several ways to train construction personnel:

- on-the-job – learning by undertaking the work through an apprenticeship

- off-the-job – learning away from work on block release, for example, at a local college

- attendance at college – taking various qualifications alongside work

- distance learning – learning not by attending college, but by working through guided tuition, for example, a course online

- open learning – distance learning courses with no set timeframe that you study at your own pace.

Accredited qualifications

There is a wide range of accredited qualifications that can be worked towards:

- Apprenticeships – work-based qualifications that can be started at age 16

- Diplomas – a suite of Foundation, Higher and Advanced qualifications for learners aged 14–19

- certificates – part-time qualifications taken while working

- degrees – full-time or part-time university-awarded qualifications

- professional qualifications – the exams required to gain full professional status

- CPD – the ongoing lifelong learning associated with your professional qualification status

- short courses relating to new developments – for example, Construction (Design and Management) regulations (CDM) 2007

- licences to practise – for example, Construction Skills Certificate Scheme (CSCS) cards, Gas Safe membership.

PLTS

Independent enquirer evaluate the information you find on professional websites

Functional skills

English use your reading skills to compare and select the best information

:BTEC Assessment activity 1.5 P6 M3

You have been asked by Connexions to help put together an information and guidance pamphlet on careers in construction as you have been working for a construction firm for over five years.

1 Identify the qualifications, training and development needed to support an architect, contracts manager and technician in the construction industry. **P6**

2 Explain how operatives and craftspeople can develop their careers and progress to a technical or professional role in a given area of activity. **M3**

Grading tips

1 To help you achieve **P6** you could visit the RIBA and CIOB websites where you will find information on the education and qualifications that are required to become a professional member.

2 To achieve **M3** you could consider the various qualifications and forms of training that can support progression.

Training on the job through apprenticeship schemes is a common route into jobs in the construction industry. Which route would you choose?

Simon Cornish
Joiner

Simon is a joiner who works with a local construction company. He started work at 16 as a modern apprentice straight from school, spending four days at work under the guidance of a qualified joiner and one day at the local college to complete his NVQ Level 2.

Simon now has six years of site experience and is ready to step up the career ladder into running construction sites and to take on the responsibilities that come with such supervision.

Simon has discussed this with his boss at work, who has said that he would support Simon with this career progression, but Simon must go and seek advice and guidance on how to accomplish this.

Simon's first call should be to a local college to seek information on the Level 3 supervisory qualification, which is the next step on from the Level 2 qualification that he has.

There are two ways of doing this, either an NVQ Level 3 or a National Certificate in construction. Either will give him the necessary skills as he climbs the management ladder. From this he can progress onto a CIOB site manager course, an HNC Construction course or a FD Construction course, and eventually complete a degree if he wishes to take it that far.

Simon needs to talk to his boss about taking on some additional responsibility at work so he can get used to running small projects and then develop experience to take on much larger projects.

Simon will need to take on some health and safety responsibilities, the control of drawings, specifications and instructions and labour management on site. He must get used to issuing instructions, ordering materials and plant and ensuring that work progresses as agreed with the client. Communication is the most important skill that he must develop in order to assist with his work as a supervisor on site.

Think about it!

1 How long will it be before Simon can be qualified?
2 What other routes could Simon take to develop his career?
3 Where can he get more information from?

Just checking

1 What is surveying?

2 What type of work can local councils issue?

3 Why are some inner city areas in need of regeneration?

4 What are the social benefits of construction?

5 What does GDP mean?

6 What does 'added value' mean in relation to a construction project?

7 What is the difference between a flat management structure and a top-down management structure?

8 What is a lateral relationship?

9 What does 'professional' mean in terms of construction job roles?

10 List four benefits of joining a professional association.

edexcel

Assignment tips

- Use the internet to research professional bodies.

- If you're interested in a career in the construction industry, find out about the level 3 construction courses available in your area. What sort of careers could they lead to?

- Find out about any local regeneration projects. What were the aims of each project?

- If you go on a site visit or if a site manager comes to your school/college:

  - research the company beforehand

  - prepare a list of questions to ask the people that you might meet, for example, you might ask the manager how they started in construction, or you might ask a general operative what they plan to do next

  - try to note down as much information as you can about the activities you see happening on site

  - ask the site manager if there is any literature you can take away.

2 Exploring health, safety and welfare in construction

Constructing buildings has its risks. Falling while working at height accounts for the greatest number of fatalities in the industry; the second greatest danger in construction is from moving machinery and plant.

The European Parliament and the UK Government have put legislation into place, covering all companies working in the industry, to try to reduce the number of people who are seriously injured or killed each year.

Every employee, visitor, self-employed worker, subcontractor or employer has a duty regarding the safety of anyone on a construction site who might be affected by their work.

Health and safety is very important. You must take into account all the risks and hazards associated with construction activities when you are working on site. The risks must be assessed and reduced, so that they are acceptable and will not cause injury to workers during operations. In this unit you will explore these aspects of health and safety and their control.

Learning outcomes

After completing this unit, you should:

1 know the importance of health, safety and welfare in the construction and built environment sector

2 be able to carry out risk assessments

3 understand the importance of control measures in risk assessment.

Assessment and grading criteria

This table shows you what you must do in order to achieve a pass, merit or distinction grade, and where you can find activities in this book to help you.

To achieve a **pass** grade the evidence must show that you are able to:	To achieve a **merit** grade the evidence must show that, in addition to the pass criteria, you are able to:	To achieve a **distinction** grade the evidence must show that, in addition to the pass and merit criteria, you are able to:
P1 outline key methods used to ensure good standards of health and safety on a construction site **See Assessment activity 2.1, page 28**	**M1** describe how human and workplace factors affect hazards and risks on construction sites **See Assessment activity 2.3, page 31**	**D1** analyse how changes in work methods affect hazards and risks on construction sites **See Assessment activity 2.3, page 31**
P2 identify the roles and responsibilities of relevant personnel **See Assessment activity 2.1, page 28**		
P3 identify potential risks and hazards in an area of the working environment **See Assessment activity 2.2, page 29**		
P4 conduct a risk assessment **See Assessment activity 2.4, page 32**	**M2** relate the findings of the risk assessment to the recommended control measures **See Assessment activity 2.4, page 32**	**D2** evaluate the impact of the risk assessment on employees, visitors to sites and the public **See Assessment activity 2.4, page 32**
P5 explain how control measures are used in risk assessment procedures **See Assessment activity 2.5, page 40**		

How you will be assessed

This unit will be assessed by an internal assignment that will be designed and marked by the staff at your centre. Your assessment could be in the form of:

- presentations
- case studies
- practical tasks
- written assignments.

Stephen, 16 years old

I had to take this unit as part of some additional learning for my Brickwork Diploma at Level 2. I thought that health and safety just meant stuff like wearing gloves while mixing cement. I didn't realise that it was so involved with laws, risk assessments, controls, reviews and action plans – all of which mean I've got responsibilities while I'm working.

This unit has really opened my eyes to just what is really involved in running a construction site, and to the health and safety responsibilities that go with this, so that I can do my bit to prevent accidents and injuries to other workers – and myself!

I enjoyed learning about what a hazard was and how it can become a risk, which then has to be controlled to prevent an accident occurring on site.
I also now know what my responsibilities will be on site, not just for myself but towards others too. This unit will make me place safety first, before anything else when I start work on site.

Over to you!

- Why do you think health and safety might be so important in construction?
- What health and safety measures do you know about already?

1. Health, safety and welfare in the construction and built environment sector

A risky business

Imagine you have designed a shopping complex for a major retail outlet and are now ready to start the project. There will be hazards not only for the workers constructing the complex, but also for the public who will use it.

What hazards must you identify in order to protect the members of the public who will eventually use your shopping complex?

1.1 Legal responsibilities

The Health and Safety at Work etc. Act 1974 (HASAWA) is the main law used in the UK to enforce safety on construction sites. Everyone on site from employees to supervisors, contracts managers to visitors, subcontractors to employers has duties specified under this Act.

Employees have a duty to:

- take reasonable care of themselves and others
- cooperate with any employer's duty.

Employers have to:

- provide and maintain safe plant and safe systems of work
- provide arrangements for ensuring safe means of handling, use, storage and transport of articles and substances
- provide information, instruction, training and supervision
- provide a safe place of work with safe entry and exit to that workplace
- provide and maintain a safe working environment with adequate welfare facilities.

The Construction Design and Management Regulations 2007 (CDM) place duties on:

- clients
- designers
- the CDM coordinator
- the principal contractor.

Did you know?

HASAWA is the only law with an 'etc.' in its title, but not many people remember to use this!

All of these people must be involved in assessing the risks involved with the design and construction of any project that is notifiable to the Health and Safety Executive (HSE).

1.2 Workplace health and safety

A workplace has to have a health and safety policy. In general terms, this is a statement from the company, laying out its organisation of health and safety and how this will be applied and adhered to.

A system of work must be in operation at all times, and is generally designed by supervisors and managers. For example, there may have to be an automatic locked gate to prevent people and machinery mixing. At points like these, where the hazards of working are identified, a risk assessment is needed (see page 31).

Health and safety management system

Management systems are used to control health and safety: for example, a no smoking policy in a chemical factory. These have to be organised by suitably qualified personnel who will implement, monitor and review the systems to ensure that they are working adequately. Employees will often sit on health and safety committees, as their first-hand experience and knowledge is essential.

Active and reactive monitoring techniques

It is no use implementing a safety system and then leaving it and forgetting it. It must be actively monitored, evaluated, and reviewed. Why? Because processes change, new materials are developed, and people move jobs. These changes have to be monitored continually for fresh risks. This monitoring can be done through:

- safety inspections
- safety tours
- safety committees
- hazard checklists
- accident investigations and feedback.

HS(G) 65 is one important guide that you may come across, which outlines best practise for implementing and monitoring safety systems in the workplace.

Reactive monitoring techniques involve dealing with potential hazards immediately, before they create risks (for example, removing rubbish from a fire exit and disposing of it quickly), as well as reporting and investigation any incidents and accidents that do occur, to try to find out why they happened and make sure they don't happen again.

All this will contribute to having a safe system of work that keeps all personnel informed of the risks and protected from injury or harm.

Did you know?

Designers have to assess the risks associated with constructing a building and with running the building during its lifetime.

Key term

Monitoring – the continual measurement of a system.

Did you know?

HS(G) 65, BS8800 and ISO18001 are all safety management systems.

1.3 Legal requirements

The Government has empowered the Health and Safety Executive to monitor safety on construction sites through inspections. If an inspector finds a safety fault, he or she can issue an improvement or prohibition notice, which can be **enforced** by law. If an employer is found to be at fault then prosecutions under HASAWA can be undertaken. This may lead to a heavy fine or imprisonment for serious breaches of the regulations.

Did you know?

The HSE has more powers than the police.

PLTS

Independent enquirer look at health and safety issues from different perspectives

Functional skills

ICT use the internet to find out about people's responsibilities for health and safety

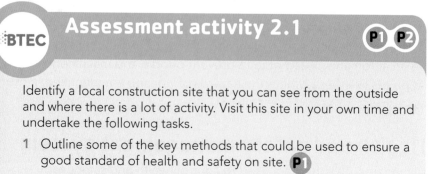

BTEC **Assessment activity 2.1** P1 P2

Identify a local construction site that you can see from the outside and where there is a lot of activity. Visit this site in your own time and undertake the following tasks.

1 Outline some of the key methods that could be used to ensure a good standard of health and safety on site. **P1**

2 Identify the roles and responsibilities of these site personnel: a site manager, an employee and a subcontractor. **P2**

Grading tips

1 To achieve **P1** you could observe a typical construction site locally, looking at the safety measures that have been put into place to ensure a good standard on site. Take a camera with you so you can study the images further.

2 To achieve **P2** you will need to look at the employer, employee and non-employee responsibilities under HASAWA to help you put into words how each of these people should behave.

2. Be able to carry out risk assessments

2.1 Hazards

A hazard is something that has the potential to cause harm: for example, electricity has the potential to give someone an electric shock when they touch a live wire. Hazards are often classified as:

- physical
- environmental
- chemical
- biological
- psychosocial.

Hazards that can exist quite happily without causing any harm to a person are said to be a low risk: for example, an electric cable on a pylon that no one can reach unaided.

2.2 Risks

A risk is the level of danger a hazard poses, and how likely this is to occur. A risk is normally classified as low, medium or high – this is known as a risk **rating**. The **severity** of the risk must be assessed, so that 'control measures' can be applied to reduce the severity of the risk to an acceptable level, in an effort to make sure that the hazard will not cause harm or injury.

Being able to spot risks comes from a thorough knowledge and understanding of the different aspects of construction operations involving plant, equipment, machinery and materials.

Key terms

Rating – a measure of how severe a hazard has the potential to cause harm.

Severity – the extent to which something is bad, serious or unpleasant.

BTEC Assessment activity 2.2 **P3**

Arrange to visit a workshop or area that you are not familiar with in the teaching environment where you are studying. Make sure that you have permission to do this – otherwise you could create a hazard just by being there!

In this environment, identify three potential hazards and explain the risk of injury they pose. **P3**

Grading tip

In order to achieve **P3** you will just need to identify three potential environmental hazards, such as a cable that is on the floor. A photograph will give you time to look for potential hazards you may not have seen at first.

PLTS

Independent enquirer explore hazards in an area you are not familiar with

Functional skills

English use your writing skills to explain the hazards clearly

Why is a crane needed on site when constructing a timber-framed house?

2.3 Work methods

A method of working, such as constructing houses in brickwork using cement mortar, is often established over a period of time. However, changes in technology can bring different work methods: for example, timber-framed housing, where no brickwork may be required.

Any change in a working method will need to be reassessed for the associated hazards and risks: for example, with timber-framed housing, a crane is needed to lift the panels into position, whereas with brickwork, no crane is needed.

2.4 Workplace changes

Many changes can occur within a workplace over time; these can be rapid or gradual, clear or barely noticeable. Workplace changes include changes in:

- temperature
- dust
- humidity
- confined spaces
- traffic access and **egress**.

Changes like these mean that potential hazards need reviewing. For example, high temperatures can lead to people fainting, getting dehydrated and even falling unconscious.

Workplace changes should be carefully monitored. Remote sensing systems are often used for this: for example, a gas test can be lowered into a confined space and an alarm bleeps if gas is present. This monitoring can be automatic, with a mechanism in place to shut down a piece of equipment if necessary, to reduce the risk.

2.5 Human factors

Fig. 2.1 illustrates the human factors that affect potential hazards and accidents for each of us in a working environment. For example, you would not put a 16-year-old apprentice in a high-risk environment for which they have no experience or training – it would be just too dangerous; instead, a person who had experience, knowledge and understanding, as well as the right attitude, would be ideal.

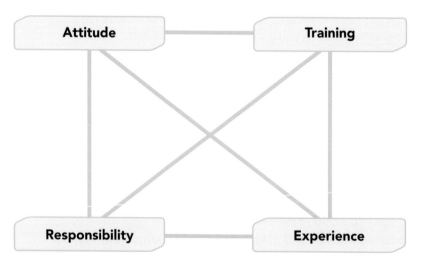

Fig. 2.1: Human factors affecting the work environment

BTEC Assessment activity 2.3 **M** **D**

Some new homes are going to be built on a construction site. Rather than using traditional methods, plans have been changed to use timber-framed construction instead for these new houses. You have been asked to undertake a risk assessment on a medium-sized housing construction site. This is a busy site with several new craftspeople working each day on different aspects of each house.

1 Describe how human and workplace factors affect the hazards and risks on a construction site. **M**

2 Analyse how the change in construction methods for the new homes will affect the hazards and potential risks. **D**

Grading tips

1 To achieve **M** you need to look closely at what a human factor is, how human factors interact with the construction environment, and how they affect types of hazard and associated risks. Relate everything to a construction site scenario or an actual example.

2 To achieve **D** you need to undertake some in-depth analysis on what the change from traditional to timber-framed construction involves. What effect will this have? What are the changes in hazards and risks? Answer these questions as fully as you can.

PLTS

Team worker work with others to identify the hazards

Functional skills

ICT use the internet to research timber-framed construction

Fig. 2.2: The hierarchy of control

2.6 Risk assessments

Fig. 2.2 illustrates the sequences you must follow to reduce the effect of a hazard. This list works as a set of steps with landings at each step; you can get off the steps at any point when you have taken sufficient measures to reduce the hazard's risk and its potential to cause harm. Always remember to monitor the environment for changes and review the control measures that you have in place already.

Fig. 2.3, page 32 is an example of a simple risk assessment form and can be used as a template.

Risk assessment form

Location _____

Activity _____

Date _____ Assessor _____

Hazard identified	Risk level low, medium or high with explanation	Control measures required
Monitoring procedures		

Fig. 2.3: A simple risk assessment form

Did you know?

PPE should always be the last resort in reducing a risk, not the first.

PLTS

Reflective learner use what you learn from your risk assessment for your evaluation

Functional skills

English use your writing skills to record information in the table

Assessment activity 2.4   P4 M2 D2

BTEC

1 Using the form in Fig. 2.3, undertake a small risk assessment of an area of work activity, identifying and recording three workplace hazards from your observations. **P4**

2 When you have undertaken this assessment, identify and describe the control measures required to reduce each hazard to an acceptable level. **M2**

3 Evaluate the impact the hazards you have identified could have on a visitor to this workplace or an employee in the environment you selected. **D2**

Grading tips

1 To achieve **P4** you will need to identify three hazards in the workplace, recording these in a suitable format, such as a small form that you have created. Remember that a hazard is something that has the potential to cause harm: for example, electricity within a cable.

2 To achieve **M2** you will need to relate a control measure, which is something that is used to reduce the risk of the hazard from causing harm, to the three identified hazards.

3 To achieve **D2** you will need to analyse and assess the impact each hazard could have on a visitor to this workplace, what could be the potential risk, what controls are required, how they will work, what needs reviewing, and so on.

3. The importance of control measures in risk assessment

3.1 Health and safety training in the construction industry

Training is essential to inform people of potential hazards and risks in their workplace, and to enable them to handle them safely. Many training routes are available, including:

- toolbox talks
- Construction Skills Certification Scheme (CSCS)
- Construction Plant Competence Scheme (CSPS)
- Construction Industry Training Board (CITB)
- Chartered Institute of Building (CIOB).

Training can inform people about new technology, highlight the risks with specialist plant, outline supervisor's responsibilities and provide a nationally recognised qualification.

A toolbox talk is the simplest route. This just consists of stopping work and gathering employees around to discuss a particular safety issue in detail so that everyone can listen, understand and contribute to the safety measures required.

3.2 Procedures

Safe system of work

A safe system of work should always be documented where possible. This can then be a useful basis for training new employees. The document should be based around a company's health and safety policy, as this is how the company's arrangements for health and safety are implemented. Any safe system of work – and the document that explains it – should be revised regularly.

Safety instructions

Safety notices and instructions are safe procedures, as are site inductions, but you must make sure that they are clear and easy to understand, for all workers.

Method statements

Method statements offer another way of recording a safe system of work. A method statement details how a particular piece of work will be carried out and what will be used to do this. It would include, for example, the method of construction, how many people would be doing the work and what plant, materials and equipment are to be used.

'Permit to work' systems

A permit to work is a useful safety tool. It lists all the measures required when undertaking a high-risk activity. Permits to work would be used for:

* hot working

* working at height

* working on roofs

* working in confined spaces.

Several signatures are often required on these forms, including those of the people doing the work, the supervisor and the client or their representative.

3.3 Protective equipment

Typical PPE used on site would be:

* hard hat

* overalls

* safety boots

* safety glasses and goggles

* ear defenders

* respiratory equipment.

All of this equipment must be maintained and stored correctly to prevent any damage that would reduce its effectiveness in preventing harm. Never forget that PPE should always be the last thing given to an operative, and should only be needed if the other 'hierarchy of control' measures have not eliminated the risk (see Fig. 2.2, page 31).

All PPE should be issued free of charge to operatives. It is then the responsibility of the operatives to report any defects and get a replacement or repair.

3.4 Substances

Substances used on site are covered by the **COSHH** regulations. A risk assessment is needed to ensure that any substance is used safely and in a manner that reduces risk from exposure.

Lead and asbestos are two substances that can have a long-term effect on workers. Lead is present in old types of paint, and asbestos has been used throughout older buildings in many different formats. Both cause long-term injury to health.

Is this bricklayer wearing the right protective equipment? What sort of things would you need to look at in a risk assessment for bricklaying?

Did you know?

An employer has to provide PPE free of charge.

Key term

COSHH – Control of Substances Hazardous to Health.

COSHH steps

COSHH risk assessments have their own hierarchy of steps to be used to reduce the risk of injury.

Step 1 Assess the risks to health.

Step 2 Decide what precautions are needed for the substance.

Step 3 Prevent or adequately control exposure to the substance.

Step 4 Ensure that control measures are used and maintained by people working with the substance.

Step 5 Monitor the exposure.

Step 6 Carry out appropriate health surveillance of workers.

Step 7 Prepare plans and procedures to deal with accidents, incidents, and emergencies involving the substance.

Step 8 Ensure employees are properly informed, trained and supervised in using the substance.

Asbestos has been found to be very dangerous, but only relatively recently; it was used in many construction products up to the 1960s. The Control of Asbestos at Work Regulations were brought in to deal with the risk. You have a legal duty to identify any asbestos within buildings that you own, and to label it as such; these details are then kept in an asbestos register. Asbestos should be removed and disposed off safely as each building becomes vacant or is refurbished, or **encapsulated** where it cannot be removed.

Key term

Encapsulate – cover or enclose completely without disturbing.

3.5 Fire precautions

For a fire to start, three things need to be present (see Fig. 2.4):

- a heat source to ignite the fire
- fuel to feed the fire, in the form of combustible materials
- oxygen for the process of combustion.

Fig. 2.4: The fire triangle

There are many different types of fire, and not all of them can be put out with water – with an electrical fire, this would be extremely dangerous. It is vital that you use the correct extinguisher to fight the classification of the fire.

Table 2.1: Fire extinguisher types and uses

Fire extinguisher type	Type of fire				
	Solids (e.g. paper or wood)	Electrical equipment	Flammable liquids	Flammable gases	Cooking oils
Water	✓	✗	✗	✗	✗
Foam	✓	✗	✓	✗	✓
Carbon dioxide (CO_2)	✗	✓	✓	✗	✓
Dry powder	✓	✓	✓	✓	✗

Did you know?

By law, you are not allowed to use a fire extinguisher unless you have been trained to do so.

Activity: Fire classification

Complete the right-hand column of Table 2.2 to show the UK and European classifications of fire. Note that no 'E' classification exists in the UK.

Table 2.2: Fire classification

Class	Type of fire
A	
B	
C	
D	
E – not a class in the UK	
F	

3.6 Good housekeeping

Keeping a workplace tidy and clean is essential to reduce hazards from objects that could cause fires, slips, trips and falls. Good housekeeping involves having:

- sufficient space for working
- safe storage of materials
- clear routes and maintained fire exits
- rubbish stations
- level working surfaces.

This reduces the amount of materials and waste around a working area, preventing many of the tripping accidents that can result in soft tissue injuries or bone fractures.

3.7 Working at height

Falling from height causes nearly 50 fatalities on construction sites every year. This is an area that the European Parliament has tried to address through the **WAH** Regulations 2005.

The WAH Regulations clearly stipulate that no work shall be undertaken at height unless it has been assessed by a risk assessment and there is no alternative to working at height. The regulations require higher safety standards than before. For example, to reach a scaffold, climbing ladders would now be superseded by a stairway platform that has handrails.

Many control measures can be used to eliminate falls from height, including:

- redesigned fixtures that lower to ground level
- fittings that can be cleaned from the inside
- mobile elevated platforms
- harnesses
- airbags
- safety netting.

Any roof work has the potential for an accident from falling. Fragile, older roofs are particular areas for caution. Older roofs may have skylights that cannot be seen as they have become dirty, or the surface may have become brittle so that, if you tread on it, it may give way.

Key term

WAH – working at height.

What are the dangers when excavating trenches below ground level?

3.8 Working below ground

When working below ground, you must consider carefully the support required for the sides of any excavation. You can simply overdig an excavation to the **angle of repose** of the soil, so that it supports itself. However, often there is not sufficient room to do this, so you will need to dig trenches or large earthwork excavations instead.

These require support using one of the following methods:

- wooden boards and struts
- steel sheet piles
- mass fill foundations.

When working in a trench, the risk comes from collapse and crush injuries, and drowning if there is any water present and you become trapped. Excavations are normally used to install foundations, drainage and services below ground level.

3.9 Confined spaces

A confined space can be classified as a space where there is no normal access or egress, such as a door. A space like this, such as a manhole, service duct or ceiling void, would need hatches or lids in order to provide access to the workplace.

The Confined Space Regulations 2001 cover working in such environments. Undertaking of a risk assessment is paramount, including looking at arrangements to recover a person who may be in trouble within any confined space – sending in someone to rescue a worker in trouble may put the second person in danger too. Rescue arrangements may involve the use of **BA** (breathing apparatus); gas tests of the atmosphere are often taken to ensure that it is not explosive and is breathable without equipment.

Control measures that may have to be used when working in confined spaces include:

- rescue equipment
- two-way communication
- gas test
- harnesses and special PPE
- ventilation systems
- special training of operatives.

3.10 Safety signs

Table 2.3 illustrates the colour coding associated with warning, prohibition, mandatory and advisory safety signs. These colours have their own meaning: for example, red normally acts as an 'alarm', such as a 'stop' traffic light, yellow signs give advice, green signs give information, and blue signs are mandatory – they must be obeyed.

Table 2.3: Safety sign colouring

Sign colour	Sign example	Sign meaning
Red	No smoking — It is against the law to smoke in these premises except in a designated area	**Prohibition** – alerting you to dangerous behaviour, or how to stop something: e.g. a tool
Yellow		**Advisory** – alerting you to take particular precautions or to be aware of a risk
Green	Fire assembly point	**Informative** – instructing you when an area is safe or where exits, escape routes or equipment are: e.g. a fire assembly point or exit sign
Blue	Hearing protection must be worn	**Mandatory** – alerting you of actions that must be taken for safety: e.g. wearing correct protective equipment

3.11 Plant, equipment and machinery

The Provision and Use of Work Equipment Regulations 1998 mean that all equipment must be inspected, maintained and tested to keep it at a safe standard, fit for its intended purpose. Operatives need to be trained so that they are **competent** to use the equipment safely.

The Lifting Operations and Lifting Equipment Regulations (LOLER) 1998 cover the use of equipment such as forklift trucks, cranes and mobile access platforms. These regulations ensure that, for example, a crane has the correct test certificates and is rated to lift only a certain amount. Records of all inspections have to be kept.

3.12 Electricity and buried or overhead services

Electricity on a construction site has to be reduced in voltage so its potential to cause harm is reduced. The normal voltage in your home would be 230V, but the voltage used on a construction site is 110V. Because of this difference, specialist equipment must be used which can accept this lower voltage. Transformers are normally used on construction sites to reduce 230V down to 110V; these only have a short lead, with a 230V plug on them.

Key term

Competent – having the qualifications, training, knowledge and experience to be able to do something.

Electricity cables are normally detected using a CAT – a cable avoidance tool – which makes a sound when it detects a cable buried in the ground. Doing this is essential when you undertake an excavation. Every cable found must be treated as live until the respective authority has certified it as dead.

Overhead cables are easy to identify, but height restrictors are needed on poles to warn of their presence: any large machinery travelling below them will collide with the pole, and not the cable, which should be a safe distance away.

Activity: Voltage colours

Copy out Table 2.4 and fill in the colour that identifies each of these voltages.

Table 2.4: Voltage colours

Voltage	Colour
110V	
230V	
415V	

PLTS

Team worker work with others to identify control measures

Independent enquirer work out what the right control measures are

Functional skills

English use your writing skills to explain how a control measure can be used

Assessment activity 2.5   **P5**

You have been given a risk assessment form to complete for the demolition of a domestic garage by your company, as the client requires a new brick one to be built. You see the words 'control measure' on the form.

Explain what a control measure is and how it can be used in a risk assessment procedure. **P5**

Grading tip

In order to achieve **P5** you will need to undertake some research on what a control measure is and how it works: for example, what would placing fencing around the site do, and what would it be a control measure for?

Josh Freeland

Health and Safety Manager

Josh completed his qualification in brickwork at Level 2, and was lucky in obtaining an apprenticeship in brickwork with a national company. He was enrolled on an NVQ Level 3 programme for the first year, and passed this successfully.

During this period of training, Josh was asked several times by the older bricklayers to complete the risk assessments on the brickwork activities they were undertaking on various contracts.

He really enjoyed breaking down the jobs into hazards and undertaking the risk evaluation, recording all of his findings on the company's standard report.

Josh had built up quite a collection of these, and had even started to undertake some method statements, which the company required for some of the specialist contracts involving permits to work.

The level of care Josh was taking had come to the attention of one of the company's health and safety officers. He was so impressed that, at Josh's next review, he approached him for the position of trainee Health and Safety Manager. Josh did not realise that the work he had produced had been noticed, and is now seriously considering this move into a valuable professional career in health and safety management.

Josh's progress is a direct result of the effort that he has put in which had become noticed and now he has an opportunity to progress within the company, receiving a career grade salary with a company car for the site inspections.

Think about it!

- **Would you take the position that is being offered?**
- **What would be the benefits of doing this – and the disadvantages?**
- **Would it be a long-term commitment?**

Just checking

1 What does CDM stand for?
2 As an employee, do you have any duties under HASAWA 1974?
3 Your employer has charged you for the safety boots you have been issued with. Comment on this.
4 Define COSHH and outline what it covers.
5 What precautions can be taken to prevent falls?
6 What should a safe system of work be based around?
7 What three things are required for a fire to take hold?
8 Name four different types of fire extinguisher.
9 What precautions are required for working in a confined space?
10 Is a red safety sign mandatory?
11 What colour is an advisory safety sign?
12 What voltage should be used on a construction site?

edexcel

Assignment tips

- Be sure to show you understand that health and safety is not just a matter of common sense. Explain that many factors are involved with an accident, including human, physical and psychological factors.

- Take a camera with you on any site visits. You can look over the photographs and spot hazards that might not have been apparent at the time, and use the photos as evidence for your assessment.

- Look for websites that can help you break down health and safety into manageable pieces that make sense, making sure you remember to acknowledge your sources.

- Ask your tutor for a copy of the risk assessment form in the tutor guide, so that you understand the pass criteria.

3 Sustainability in the construction industry

Global climate change presents a challenge to us all. The earth is experiencing a global rise in temperature, which may be due to human activity. The extensive use of fossil fuels has released carbon dioxide (CO_2) into the atmosphere, trapping heat and making the temperature rise.

Sustainable design and construction can help to counter the release of CO_2, by stopping it escaping at all, or through using materials that lock CO_2 into their structure.

Sustainable methods can also help save other of the earth's finite resources, such as supplies of oil and coal, so that they can be there to meet the needs of future generations. Once resources like these have been consumed, they can never be replaced.

Through minimising waste, controlling pollution, and using sustainable construction methods, all those in the construction industry can – and must – play their part in protecting our natural environment.

Learning outcomes

After completing this unit, you should:

1   understand the concept of sustainability as it applies to the construction and built environment sector

2   know the issues affecting the development of a sustainable built environment

3   know how sustainability can benefit the built environment both locally and nationally

4   know how sustainable design and construction techniques are used to address environmental issues.

Assessment and grading criteria

This table shows you what you must do in order to achieve a pass, merit or distinction grade, and where you can find activities in this book to help you.

To achieve a **pass** grade the evidence must show that you are able to:	To achieve a **merit** grade the evidence must show that, in addition to the pass criteria, you are able to:	To achieve a **distinction** grade the evidence must show that, in addition to the pass and merit criteria, you are able to:
P1 explain what is meant by sustainability **See Assessment activity 3.1, page 48**	**M1** assess the benefits of considering sustainability issues in the built environment **See Assessment activity 3.2, page 50**	**D1** evaluate the consequences of not considering sustainability issues in the built environment **See Assessment activity 3.2, page 50**
P2 explain the relevance of sustainability to the construction and built environment sector **See Assessment activity 3.2, page 50**		
P3 identify the issues associated with the provision of a sustainable built environment **See Assessment activity 3.3, page 52**		
P4 describe the issues associated with the provision of a sustainable built environment **See Assessment activity 3.3, page 52**		
P5 identify the benefits of using sustainable construction, in both local and national terms **See Assessment activity 3.4, page 60**	**M2** compare the local and national benefits of sustainable construction in social and economic terms **See Assessment activity 3.4, page 60**	**D2** justify the selection of specified sustainable construction in terms of effectiveness and relative cost **See Assessment activity 3.4, page 60**
P6 describe the benefits of using sustainable construction, in both local and national terms **See Assessment activity 3.4, page 60**		
P7 identify the sustainable design and construction techniques used to minimise environmental impact **See Assessment activity 3.5, page 60**	**M3** evaluate the effectiveness of sustainable construction techniques at each stage of the development process **See Assessment activity 3.5, page 60**	
P8 describe the sustainable design and construction techniques used to minimise environmental impact **See Assessment activity 3.5, page 60**		

How you will be assessed

This unit will be assessed by an internal assignment that will be designed and marked by the staff at your centre. Your assessment could be in the form of:

* presentations
* case studies
* practical tasks
* written assignments.

Sam, 16 years old

Before I started this unit, I thought that sustainability was just saving coal, oil and natural gas – I knew that these are finite, and can't be replaced once they are gone – but I didn't realise that it involved much more, at a local level as well as globally.

I've learnt that sustainability involves lots of different techniques, which we can use to conserve our natural resources – things like sourcing materials locally to reduce the distance for transport, using recyclable materials and putting bus routes through new housing estates to reduce car use. You have to think about every aspect of construction that can save energy, resources and future maintenance costs.

There's a sustainable building in my local park, with the warden's house, a café and toilets. It's made entirely of cedar timber, a material that grows, soaks up CO_2 and doesn't need maintenance. I reckon this unit will have quite an impact on my attitude when I start work on a construction site. I'll be looking out for ways to work in the most sustainable way possible.

Over to you!

* What do you know about sustainability?
* What areas might be involved in 'sustainable construction'?
* Why do you think sustainability might be important?

1. Sustainability, construction and the built environment

Build up

Timber!

Have a look at the timber in your local DIY store. Does it come from a local source? Does it come from a managed forest? Will it need treating? Is it labelled as being sustainable in any way? Is all the timber softwood?

What materials might be used to construct a sustainable building?

Key terms

Finite – things that cannot be replaced once used.

Fossil – a natural carbon-based material.

1.1 Definitions of sustainability

What does sustainability mean? In practice, there are several different ways to look at it.

- Social sustainability means having green, open spaces for communities in dense urban areas, and facilities for tenants and owners on housing estates to mix and interact, such as playgrounds, sports facilities and pedestrian walkways.

- Physical sustainability covers physical acts on a construction site. For example, you could improve physical sustainability by using smaller components that can be lifted by hand, rather than with a crane, or by using modular components that can be manufactured off site then assembled on site more quickly and efficiently, using fewer resources.

- Economic sustainability looks at the costs across the whole lifespan of a project from design, through construction, and maintenance, to eventual demolition. Sustainable economic benefits can come from, for example, spending more on higher quality materials that save long-term maintenance costs, or considering the running costs of the building at the design stage.

1.2 Relevance of sustainability

Finite resources and shortages

The earth has precious **finite** resources, on its surface and underground, including the main **fossil** fuels we use: oil, natural gas and coal.

Most of our consumer economy relies on these resources to manufacture and provide energy for the goods and services our modern world demands. Our fossil fuels will eventually run out, causing shortages and price rises as reserves dwindle. New technology and renewable energy is slowly been developed, to reduce our reliance on fossil fuels and make them last longer.

Global warming

The earth is warming up. Here are some possible reasons that have been put forward.

- The orbit of the earth about the sun is not a true circle, but deviates: the nearer we are to the sun, the warmer it is.

- The earth wobbles on its axis, causing north and south to receive more heat from the sun.

- Burning fossil fuels releases CO_2, which traps heat within the atmosphere.

What are the consequences of this warming of the earth's climate?

Melting ice-caps
Ice in the glaciers and land-based ice shelves of Greenland and the Antarctic are melting and receding. The water created flows into the sea, causing a rise in world sea levels.

Rising sea levels
Sea levels are rising slowly. In the UK, it is predicted that the Thames barrier will only protect London until 2070. Rising sea levels coupled with storm winds can cause serious damage to our coastlines by eroding away the landmass. To control this, we spend millions of pounds on sea defences every year.

Climate change
Our climate is changing. Within just two years recently, the UK experienced two 'once in hundred years' floods, each resulting in huge damage to property. Our summers seem wetter, and our winters warmer; this has an effect on plants, trees and other species that rely on the seasons to control their growth and development, such as deciduous trees.

Flooding
Flooding can have a devastating affect on our buildings, and our communities. Floods can be caused by:

- lack of maintenance to drainage systems
- higher than normal rainfalls, due to climate change
- large thunderstorms causing flash floods
- rising level of sea tides
- pump failures.

Extinction of species and the impact on biodiversity
While global warming is causing changes to climates across many major landmasses, this is having its own impact on **biodiversity**. Many species will need to evolve to adapt to these changes, or will become extinct. As species become extinct, the natural order of nature will be upset. For example, the loss of predators could cause infestations of locusts or mice.

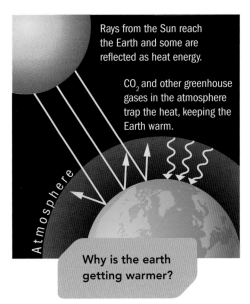

Rays from the Sun reach the Earth and some are reflected as heat energy.

CO_2 and other greenhouse gases in the atmosphere trap the heat, keeping the Earth warm.

Atmosphere

Why is the earth getting warmer?

Did you know?

The period for May to July 2007 was the wettest on record for 250 years.

Key term

Biodiversity – the range of plants and wildlife within a region.

Key term

Carbon quota – the amount of CO_2 (measured in tonnes) that a country can emit from industrial processes under international agreements.

We have to conserve the finite resources of the earth for future generations, to ensure their survival. This means we must save energy, and act sustainably, carefully managing our existing resources so they will last longer.

Local and global context, and the impact on construction

If we include sustainability locally, in every part of our daily lives, this will have a profound affect globally.

Currently, less developed countries may sell their **carbon quotas** to countries that are vast consumers, to balance out the level of emissions; in return, they receive technologies to help them develop. If we are to reduce the change in global temperatures, we must act together, and every country must meet its agreed targets.

As far as construction projects go, sustainability now needs to be carefully considered at every stage, from design onwards – and those working in each sector need to plan together how to make projects as sustainable as possible, throughout their lifecycles. For example, turning a building to face south will have a huge effect on heating and lighting costs.

What areas of the UK were affected by flooding in 2007?

Assessment activity 3.1

BTEC **P1**

You have just started a part-time job at a local builders merchant, helping in the yard. Clients keep using the term 'sustainability' when they place orders for materials.

Explain what is meant by 'sustainability'. **P1**

Grading tip

To achieve **P1** you will need to define what sustainability is, explaining what it means in clear, appropriate terms.

2. Developing a sustainable built environment

2.1 What are the issues for the built environment?

Nature of the built environment

People have been altering our built environment for thousands of years, expanding from villages into towns and major cities; very little natural landscape is left.

Some of our heritage is protected through the graded listing of historical buildings by English Heritage, one of the organisations that look after them for the benefit of future generations. The creation of 14 green belt areas has helped to contain our **conurbations**, and keep them linked to the natural environment.

Impact on the natural environment

We all need buildings to live in. As population levels have risen, housing developments have grown from villages to towns and cities, reaching far into the countryside. More recently, the development of out-of-town retail complexes has led to urban decline.

In the past, the tradition was to develop 'greenfield' sites, untouched areas of land that that had not yet been built on. Now the more sustainable trend is to recycle existing derelict sites, known as 'brownfield' sites. On these sites, contamination may have to be dealt with and the infrastructure improved, but great regeneration areas can be created.

Our duty to present and future generations

The construction industry must change some of its traditional ways of operating if it is to help safeguard and maintain the environment for the future generations.

Urban regeneration is one technique the industry can adopt. Through government backing and grants, inner cities can be redeveloped using sustainable modern buildings. Leeds, Glasgow and Manchester have all benefitted from urban regeneration, carried out with the environment in mind.

The home of the future

The government's long-term target is to produce a 'zero carbon' home. This would be **carbon neutral** in its construction and be carbon-free in its running costs.

Key term

Conurbation – an extensive urban area formed by growing cities merging together.

What are the benefits of green, open spaces within the built environment?

Key term

Carbon neutral – neither adding carbon to the atmosphere nor taking it away.

49

Achieving this will take a tremendous effort, and its success relies on:

- re-educating the next generation
- full government backing
- developments in energy technology
- increases in renewable energy sources
- people being willing and able to pay a premium on a house purchase.

Key term

Relevance – importance for and connection to something.

PLTS

Independent enquirer develop reasoned arguments to persuade the director

Creative thinker ask questions that extend your understanding of sustainability

Functional skills

English use your writing skills to communicate your ideas persuasively

BTEC **Assessment activity 3.2**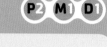

Your construction director has not taken on board any sustainable construction techniques, describing them as 'daft'. You know that this is a mistake and during a discussion you try to convince the director that she is wrong.

1 Explain the **relevance** of sustainability to the construction sector. **P2**

The director is still not convinced and has asked you to carry out some more research to prove your points.

2 Assess the benefits of considering sustainability issues for the built environment. **M1**

3 Evaluate the consequences of taking the director's line and not even considering sustainability issues for the construction industry. **D1**

Grading tips

For this whole activity, you need to look specifically at what effect the construction industry has on its environment and how sustainable approaches will substantially improve this impact.

1 To achieve **P2** you must explain by describing the relevance of sustainability (why it matters), what effect it is having on our built environment, and what those who work in the construction can do about it.

2 To achieve **M1** you will need to list some benefits of considering sustainable issues and then explain how they impact on the built environment in sustainable terms

3 To achieve **D1** you must discuss the advantages, disadvantages and consequences of not considering a sustainable construction approach, and explain what would be the effect of ignoring it.

2.2 What are the social and economic issues?

Meeting local and national needs

Sustainability must meet the needs of the community involved. For example, it is no good implementing cycle pathways too far from a city centre because people will still use their cars. Careful consideration and the involvement of the community are essential in any potential future development. Employment opportunities must be considered, both locally and nationally, as must the movement of labour around the country. Here, rail may be a greener option, as the rail system has already been modernised through electrification.

There is a shortage of housing in the UK, especially in the south. To resolve this, the government has a drive to provide 'sustainable communities' where low-cost, affordable, sustainable housing can be developed, as well as green public spaces in rural and urban areas.

Improved business and employment opportunities

Sustainability brings with it new opportunities for business, and can create jobs. Sustainable businesses are just starting to develop and prosper. From service industries through to hotel accommodation, all are taking sustainability on board and this can bring with it new opportunities. For example, the recycling industry has seen tremendous growth, as has the manufacture and installation of renewable energy sources such as wind turbines. Jobs bring in wealth, which you need to create or buy sustainable products and technology. At the same time, being more sustainable gives a business protection, helping it to grow stronger, and last longer than the competition.

Skills development

The introduction of sustainability into our built environment gives people the chance to develop new skills. For example, new renewable energy technology has created opportunities for:

- electrical engineering
- combined heat and power installation
- solar panel installation, servicing and maintenance
- small wind turbine installation and servicing
- offshore civil engineering.

This has stimulated growth in the sector, with opportunities for new apprenticeships, retraining for those made redundant from heavy industries, and many spin-off industries.

> Why do you think the construction of wind turbines is a growth industry?

> **Did you know?**
>
> In 2009 there were over 2500 wind turbines in the UK supplying electricity to 2.1 million homes, saving 400,000 tonnes of CO_2 emissions.

Positive economic impact

Becoming more sustainable usually involves an increased financial commitment at first, but all those involved in a project must now consider costs across the whole lifecycle of the project, rather than just the immediate costs.

Sustainability provides a much better, more stable environment for us and our children to grow and develop in. This will pay dividends: in the future, our children can 'reinvest' the sustainable choices we have made for them, by making sustainable choices of their own.

Sustainability also has the potential to make a significant contribution to prosperity, for individuals, communities and whole nations, adding to their GDP. Times of turmoil in the world economy show how vulnerable we can be to negative economic growth. Creating a lifestyle where quality counts for more than quantity must surely be better for the environment, and for our long-term future.

Negative social impact

Overdeveloping our environment can have negative effects for all of us, including:

- congestion from traffic
- local pollution
- increased pressure on waste disposal
- higher land values.

This negative impact must be considered alongside the benefits of any development. For example, a new shopping centre may create jobs, but the increased amount of traffic may cause severe congestion and pollution. These days, local planning authorities are starting to include sustainability as a desirable factor when considering planning applications.

Key term

GDP – gross domestic product, the amount of 'income' a nation has from what it produces and sells.

PLTS

Independent enquirer evaluate planning policies on sustainability

Functional skills

ICT use a DTP package, PowerPoint or tables to present your key points well

BTEC Assessment activity 3.3

You have submitted a planning application for a sustainable housing development. The local planning authority has asked you to clarify which aspects of the development are sustainable.

Identify and describe each of the issues associated with the provision of a sustainable built environment proposal. **P3 P4**

Grading tip

Make a list of key points that must be considered for a sustainably built environment.

To achieve **P3** you just need to identify the issues involved. This could be in the form of a simple list. To achieve **P4** you need to add a description of each of the issues associated with providing a sustainable environment. Your final list could have the issues themselves highlighted, followed by a description for each.

3. How sustainability can benefit the built environment, locally and nationally

3.1 The local benefits

Employment

Building decent homes increases demand for quality and attracts people into an area to live and work. This brings in businesses, which provide employment opportunities for local people. Employment is a key aspect of sustainability, and it must be considered when planning such developments; without it, people will have no chance to work and then reinvest in the local economy.

Social benefits, community consultation and local involvement

Creating clean, secure, sustainable houses provides social benefits for developments. Providing well-lit, open pedestrian areas makes residents feel safe at night, and reduces vandalism.

An essential principle of sustainability is getting all the people that matter involved with a development, through community consultation. Making the community feel part of the decision-making process helps create a social sense of belonging, reduces people's fears about a development, and allows end users to have a say in the end product. This brings benefits as involvement makes people more likely to look after their local environment, lowering crime and anti-social behaviour. Giving tenants and residents an opportunity to contribute can improve an area even further, through initiatives like Neighbourhood Watch schemes.

Any development will involve local involvement as part of the planning process. The local community will be asked for their views about a proposal, through notices or letters inviting comment, so that their opinions can be considered.

Various planning groups cover conservation areas and historical sites, while major committees involve public enquiries for roads, railways and other large developments. All these give the local population the chance to challenge or even change the proposals.

Green spaces

Including green space in a development is another key principle of sustainability. Having trees, other planting and open areas for activities gives a community somewhere to relax and socialise. On a larger scale, the Government has provided 14 green 'belts' around cities, and the UK has a network of national parks for the public to use and enjoy.

Did you know?

English Heritage has protected some 370,000 historic buildings.

53

Do you think this harbour redevelopment would be an attractive place to live?

Aesthetics

'Aesthetics' is about the way things look. Sustainable design often involves creating aesthetically pleasing buildings, as the emphasis is on using natural materials, such as cedar boarding. An attractive environment appeals more than one that is old and run down. Good design for homes makes an environment a pleasant place to live and work. This adds value to the environment at little initial cost.

Improved environments

The Government and the European Union commit funding to areas deemed to be 'enterprise zones'. These are areas of high unemployment, which have lost industries because of recession or market changes. By providing 'start-up' investment, they encourage many more investors to come into the development and support it.

Regeneration

Urban regeneration is the process of reusing or clearing derelict buildings from inner city areas and replacing these with new developments. Regeneration can have a profound effect on a run-down area. Many old warehouses in coastal cities such as Liverpool and Hull have been converted into housing and shops, which transform the built environment, provide local employment and attract tourists.

3.2 The national benefits

Cleaner air

Many cities, where fossil fuels are widely used and traffic levels are high, suffer from a problem called photochemical **smog**. Smog results from the build-up of exhaust particles in the atmosphere, which causes air pollution and this can cause lung disease. The global drive to reduce exhaust emissions brings the added benefit of cleaner air. In the UK the local authority must continually monitor air quality to ensure that emissions are below the levels set.

Reduction in flooding

Reducing CO_2 emissions should halt the rise of the earth's temperature, but this could take many years. The current rise in sea levels and climate-induced thunderstorms should return to normal.

Sustainable measures that can be employed to reduce flooding include:

- SUDS
- ponds and lakes
- swales and basins
- separate surface and foul drainage
- porous hard surfacing.

These all aim to reduce the run-off into watercourses, which reduces the likelihood of river levels rising above bank level, causing flooding.

Key term

Smog – fog that has become mixed and polluted with smoke.

Key term

SUDS – Sustainable Urban Drainage Systems, which aim to decrease and slow down surface run-off, or divert it for other useful purposes.

Swale – a hollow or marshy depression between ridges.

Changing education

For sustainability to be a success, current and future generations need to be re-educated. Teaching people not to leave equipment on standby, to turn down the heating, to buy a more fuel-efficient car, and to recycle everything can produce individual benefits, but if we all adopt these principles, the changes will have a national impact.

The Government has introduced sustainability topics into many curriculum syllabuses to change attitudes among younger people. Public awareness campaigns on issues such as using fewer plastic bags can play their part in creating a more sustainable future. When supported by the media, these can have quite an impact. For example, one market town banned plastic carrier bags, with retailers only serving people who brought their own bag. After this, several supermarkets began to charge customers for plastic bags to discourage their use.

Conservation of resources

Employing sustainability techniques in design and construction will:

- extend the life of our finite resources
- give us time to develop and implement alternative energy sources
- allow time for changes in technology, to reduce our reliance on oil and gas for power and fuel.

Economic well-being

Having a stable, sustainable economy brings many benefits, not only for the individual but also for the Government.

Table 3.1: The economic benefits of sustainability

The individual	The Government
Increase in ownership of own home	Increased revenue from taxation
A healthier lifestyle	Less strain on national health services
Produces a better place to live	Reduction in infrastructure congestion
Cleaner air	Reduction in crime figures

Environmental protection

Sustainability protects our environment by:

- establishing environmental laws and regulations
- reducing CO_2 emissions
- educating future generations
- creating green spaces
- using environmentally friendly materials.

These measures create an awareness of how we can harm our environment, and help to protect it for the future.

Did you know?

In 2007 some 55,000 homes were flooded; 35,000 of them had no warning.

Why do you think tourists visit sustainable developments, like the Hockerton Housing Project?

Better quality standards

Sustainability often involves spending more on better quality products and materials that will last longer into the future. The bonus associated with this is that it produces a better standard and often a more aesthetic look to a building or project.

Government benefits

For the Government, sustainability brings its own benefits, including:

- a reduction in road traffic, meaning lower road maintenance costs
- increased development of sustainable transport systems, such as trams, to service communities
- cleaner air, which leads to better health for all
- a reduction in CO_2 emissions, lessening the problems for the future.

Tourism

Eco-tourism is a new form of tourism linked to sustainability. Many sustainable projects, such as the Hockerton Housing project, now provide tours and opportunities for educational events, raising revenue that can be put back into expanding the sustainable community. You can visit the Hockerton website, which provides much more information about the way in which they live as a community.

Activity: Hockerton

Take a look at the Hockerton Housing Project website, and find out how people there live, work and interact as a community. Answer these questions as if you were considering this lifestyle.

- What would be the benefits for me as an individual of living like this?
- What life changes would I have to make?
- Would I still have the same standard of living?

Now discuss and reflect on your answers with a partner.

4. Sustainable design and construction techniques used to address environmental issues

4.1 Influencing factors

Stages of the development process

Sustainability techniques can be used at each stage of the project to address environmental issues.

- Planning can involve providing incentives for developers to incorporate sustainability, provide cycle and pedestrian zones, specifying the use of brownfield sites, support from EU regeneration grants and monitoring of emissions.
- Design can involve incorporating sustainable designs, building orientation, green roof technology, rainwater harvesting and grey water systems, sourcing and specifying local materials, using zero carbon materials and alternative energy sources.
- Construction can involve using prefabricated units, modular units and timber-framed construction, minimising waste on site and reducing pollution and landfill.

Factors influencing these stages

Many factors influence the planning, design and construction of a project, as you can see in Table 3.2.

Table 3.2: The stages of construction and their effect on the environment

Stage	Factors influencing these stages: physical, technical, financial, legal and aesthetic	Impact of each stage on the environment
Planning	Space available to develop The local plan, in which the council decides what will be built and where Greenfield or brownfield site Costs of decontamination of ground Government grants available Historical nature of the site and surroundings	Planning with regard to brownfield sites Planning with regard to greenfield sites
Design	Costs of an initial sustainable design Legal constraints through planning – height, size, etc. Physical size of the site Aesthetics – if a building looks good, it can 'lift' the area and create further opportunities	Visual impact depending on what the designer wants the building to look like Materials specified by the designer Services and energy used in the design
Construction	The Construction Design and Management Regulations 2007 Construction costs needing to be within budget Sustainable construction techniques needing to be employed Waste management from construction operations	Noise, dust and waste disposal, which can have an environmental impact on the local community Congestion of roads by construction traffic Use of non-local resources, adding to transport Use of greenfield sites

4.2 Respecting the natural environment

We must look after our natural environment and protect it. It is a valuable resource that we enjoy for recreation and pleasure: for example, mountain biking , hill walking and fishing are all activities that rely on the protection of the natural environment. On an individual level, respecting this shared resource means protecting it, by doing things such as taking home your litter and keeping to marked paths; for the construction industry, respecting the environment has many other aspects.

Minimisation of waste

Construction waste costs money; anything thrown into a skip is a wasted resource. You need to minimise the amount of waste removed from site, as a high percentage of the initial material brought to site could go unused.

Waste can be reduced by:

- ordering the correct quantities and sizes
- sorting waste into recycle skips: e.g. timber, metals
- incorporating waste into the design: e.g. earth bund landscaping
- using readily available design sizes.

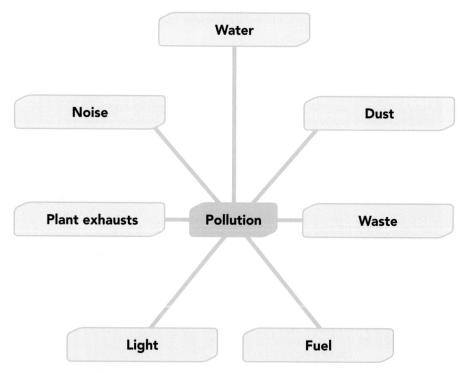

Fig. 3.2: The pollution created by construction

Reduction of pollution

Construction creates several forms of pollution, as shown in Fig. 3.2.

Careful management and planning of a construction project by qualified site managers will help reduce the amount of pollution created. A project that uses sustainable materials will reduce pollution in the long term.

Control of rate of consumption of valuable resources to conserve natural assets

Sustainable techniques use products and materials that are recyclable, or that have been manufactured from recycled materials. This slows down the rate of consumption of finite resources, as they are used over and over again.

Preservation of wildlife, flora and fauna; protection of biodiversity

Sustainability protects our biodiversity. It helps preserve nature's balance by reducing the potential harmful impact of a project, perhaps by altering a design to accommodate wildlife, or keeping the project in harmony with the natural surroundings. Sustainable projects often incorporate landscaping into the design: for example, planted roofs, indoor trees and external wildlife corridors.

4.3 Sustainable construction techniques

There are many techniques that can be used within sustainable construction. The following table gives some examples.

Table 3.3: Sustainable techniques and examples

Technique	Example
Environmentally friendly design	Using a timber-framed construction
Specifying locally sourced materials	Using locally made bricks
Improving site management and resource management	Buying locally, using local labour and contractors, prefabrication off site
Improving waste management	Sorting construction waste into recycle skips
Reclaiming and recycling	Bringing a crusher on site and recycling construction demolition into hardcore fill
Alternative energy technology	Incorporating solar panels into the roof tiles

Assessment activity 3.4

P5 P6 M2 D2

The designer of your company's housing scheme is not convinced about sustainable construction techniques.

1 Identify and describe the benefits of using sustainable techniques in both local and national terms. **P5 P6**

2 Compare the local and national benefits of sustainable construction, in social and economic terms. **M2**

3 Justify the selection of specified sustainable construction in terms of its effectiveness and relative cost. **D2**

Grading tips

Evaluate the advantages and disadvantages of sustainable construction techniques.

1 To achieve **P5** you need to identify the benefits of using sustainable techniques, which could be just a list. To meet **P6** you will need to outline why using each technique has a positive impact locally and nationally.

2 To achieve **M2** you will need to compare the benefits locally and nationally, identifing the differences in economic and social terms.

3 To achieve **D2** you need to present supporting statements on how well these methods work and how much they cost compared to how well they work.

PLTS

Independent enquirer compare and evaluate different sustainable construction techniques

Functional skills

English use your speaking skills to convince the designer

PLTS

Independent enquirer assess what the construction manager would think of the techniques you research

Functional skills

English use your writing skills to describe different techniques

Assessment activity 3.5

P7 P8 M3

The construction manager for the housing development built near your home is not convinced about the benefits of sustainable construction techniques. Answer the following questions.

1 Identify and describe the sustainable design and construction techniques used to minimise the environmental impact of a project. **P7 P8**

2 Evaluate the effectiveness of sustainable construction techniques at each stage of the development process. **M3**

Grading tips

1 Take a close look at a sustainable construction technique and evaluate the environmental benefits to the environment. To achieve **P7** you will again need to identify the techniques that could minimise the impact on the environment. To achieve **P8** you will need to give a description against each, explaining how that technique minimises the impact on the environment.

2 To achieve **M3** you will need to take each of your sustainable techniques and evaluate them in terms of whether they work or not during each stage of the development process.

Jo works in a design office for a housing contractor. This is her first job after leaving college with her National Diploma in Construction.

Jo did not realise until she went to college that sustainability was so important, or that there was so much demand for it. Jo now appreciates that designers must include this in every design they produce, as we need to save energy and resources for our future generations.

Jo's company is very keen for her to take on board this aspect and hopes to produce high-quality sustainable home for families as its main market. This is the project that Jo is assisting on. She has learned so much about the energy that is used to produce materials and what effect a project carbon footprint has on the environment, and how to include materials that have a low carbon footprint into the house design.

Jo's company is building its first 'zero carbon' show home this year, which Jo has helped design. Now she can actually see the sustainable construction techniques and specification put into a live project. Everyone is really excited about this and they all see it as the next generation home.

After several years working at the design practice, Jo has now become something of an expert on sustainable home design: she has just finished a degree in this specialist topic and is studying for a higher degree. The manager of the design office is very impressed with Jo's performance so far.

Think about it!

1 **What will be the long-term sustainable benefits of a zero carbon home?**

2 **What will it cost to achieve this?**

3 **Will sustainable homes change our climate?**

Just checking

1 What are the effects of global warming?

2 What can we do to reduce the effects?

3 What are the three aspects of sustainability?

4 What is the difference between greenfield and brownfield sites?

5 Are there any social and economic benefits in employing sustainable techniques?

6 Why would you get the local community involved within a project?

7 How can sustainability increase tourism?

8 How can sustainability protect the natural environment?

9 How can you reduce wastage on a construction site?

10 Does sustainability have national benefits for the UK?

edexcel

Assignment tips

- Find a locally sustainable construction project near where you live. You will learn a lot just by observing what is going on site, and you can use what you find out in your assessments.

- Have a look at the government's environment and greener living website and research sustainable communities in your area.

- Look at some other sustainable community websites, for example the Hockerton Housing Project, for other sources of information.

- Look at the information given by materials manufacturers to find out how sustainability is a feature of their products.

8 Exploring carpentry and joinery

Timber has been used for thousands of years as a material to provide the structure for our homes. Over the years people have learned how to refine it with detail and machining, and have worked it into many features of listed buildings.

The craftspeople who use timber are called carpenters and joiners. They assemble the floors and roof, install the windows and create the finishing details that help make our homes warm, attractive and comfortable to live in.

Timber is a natural resource that is renewable. This means once it is cut and processed we can plant and grow more to replace it. Timber is an easy material to work with, by hand or machine. In this unit, we shall explore the basic tools required to work timber as well as the health and safety aspects of using this material, and we will undertake some simple setting out and cutting of joints.

Learning outcomes

After completing this unit, you should:

1 know the hand tools and materials commonly used to perform carpentry and joinery tasks

2 understand the important health, safety and welfare issues associated with carpentry and joinery tasks

3 be able to apply safe working practices to mark out and form joints for a timber frame to a given specification.

Assessment and grading criteria

This table shows you what you must do in order to achieve a pass, merit or distinction grade, and where you can find activities in this book to help you.

To achieve a **pass** grade the evidence must show that you are able to:	To achieve a **merit** grade the evidence must show that, in addition to the pass criteria, you are able to:	To achieve a **distinction** grade the evidence must show that, in addition to the pass and merit criteria, you are able to:
P1 identify the hand tools used to perform carpentry and joinery tasks **See Assessment activity 8.1, page 69**	**M1** justify the use of hand tools and materials safely to minimise health, safety and welfare risks **See Assessment activity 8.4, page 78**	
P2 select the hand tools required to perform given carpentry and joinery tasks **See Assessment activity 8.4, page 78**		
P3 identify the materials used to perform carpentry and joinery tasks **See Assessment activity 8.2, page 70**		
P4 select the materials required to perform given carpentry and joinery tasks **See Assessment activity 8.4, page 78**		
P5 identify the PPE and safe working practices used to perform carpentry and joinery tasks **See Assessment activity 8.3, page 74.**	**M2** justify the use of appropriate PPE and safe working practices to minimise health, safety and welfare risks **See Assessment activity 8.3, page 74**	
P6 explain the selection of the PPE and safe working practices to be used in given carpentry and joinery tasks **See Assessment activity 8.3, page 74**		
P7 produce setting out rods and use them to mark out work **See Assessment activity 8.4, page 78**	**M3** produce finished work with all joints within 3 mm tolerance and square **See Assessment activity 8.4, page 78**	**D1** produce finished work with all joints within 1 mm tolerance and square **See Assessment activity 8.4, page 78**
P8 set out and cut joints in timber **See Assessment activity 8.4, page 78**		
P9 use a range of joints to produce a timber frame to a given specification **See Assessment activity 8.4, page 78**		

How you will be assessed

This unit will be assessed by an internal assignment that will be designed and marked by the staff at your centre. Your assessment could be in the form of:

- presentations
- case studies
- practical tasks
- written assignments.

Beth, 16 years old

I used to watch my Dad doing carpentry, and thought it looked really satisfying – starting with just a piece of wood and turning it into something that you could use and enjoy every day, something to be proud of.

When I started my joinery practical work, I started to realise how accurate you have to be to create a finished joint. First you have to produce a drawing of the measurements, then you set out the joint using squares and a pencil.

You have to use really sharp tools to cut out the joints or you will not get an accurate finish, so you really have to take care. So many carpenters and joiners damage their hands, even those who've been doing the job for ages.

This unit was difficult at first, as I have never done this type of work before, but you soon learn how to get it right by practising the joints before taking the assessed piece.

I really enjoyed this unit because you learned about the theory side of carpentry and joinery and then tried it out on the practical joints that we had to construct. Now I'll be using these for some much bigger projects, like putting on a roof, on my next course.

Over to you!

- What sections of this unit might you find challenging?
- Which particular section are you looking forward to most?
- What can you do to prepare yourself for the unit assessments?

1. Hand tools and materials commonly used for carpentry and joinery

Build up

The perfect fit

Take a look around the room you are in now. Identify the parts that are made from timber. Can you see any furniture, framework or units that have been made using timber joints? Each of these has had to be fitted together without the use of nails, as this would appear unsightly.

- How do you think these joints have been constructed?
- What stops them falling apart?

1.1 Hand tools

Pencil

A carpenter uses a pencil to produce a setting out rod or a template to identify the dimensions of the joint. A pencil essentially consists of a graphite core surrounded by a timber laminate; once sharpened, it produces a line on a surface when drawn across it.

Carpenter's pencils come in two styles: a hexagonal cross-section, or a elliptical (oval) cross-section. The oval carpenter's pencil is that shape to stop it blowing away in the wind when used outside.

Did you know?

You can buy a special carpenter's pencil sharpener!

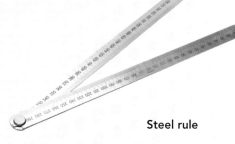

Steel rule

Steel rule

This sort of ruler will not be damaged when you use a marking knife to dimension joints or cut across a line. It is normally graduated in centimetres as well as millimetres, but only millimetres and metres are used in construction.

Combination/tri-square

This is a very useful tool that consists of a ruler with a groove in it. You can slide different squares onto this groove to produce a 90° angle, a 45° angle and an adjustable angle. Sometimes the handle of the square has a levelling bubble inside it.

A combination square

Marking knife

A marking knife consists of a sturdy handle with a sharp steel blade held within it. You can replace the blades by unscrewing the handle. Take care with this tool as a slip can cause a serious cut. Some knives have a retractable blade, to keep it inside the body of the knife when not in use.

Marking/mortise gauge

These two types of gauge are used to mark out a line parallel with a face when dragged along it. A mortise gauge has two prongs that can be adjusted to mark out a tenon and a mortise to the same dimensions, so they will fit. Each of these gauges has a steel pin that is sharpened so it can cut a line along the timber.

Sliding bevel

A sliding bevel has a blade with a slot along half its length and a handle with a screw in it, which passes through the slot in the blade. By loosening the screw you can adjust the blade to measure and mark any angle.

Wooden mallet

You use a wooden mallet with a chisel to cut into timber. Because it is wooden, it does not damage the head of the chisel, which is often made of box wood.

Claw hammer

A claw hammer gets its name from the 'claw' you can see on one end of the head. You can use this to extract nails from timber by slotting it around the nail head, using some leverage to pull back on the hamme and removing the nail.

Mortise/bevel-edged chisel

Chisels are used to remove sections of timber, for instance, to create slots or mortises for joints. A mortise and a bevel-edged chisel have different cross-sections for different uses.

Marking gauge and mortise gauge

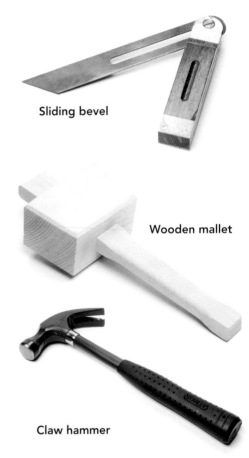

Sliding bevel

Wooden mallet

Claw hammer

Activity: Chisels

The two different types of chisel have different cross-sections. Take a look at both chisels, taking care to keep any protective caps on the ends or the blade away from you, then answer these questions.

- **Why has the mortise chisel got a rectangular section?**
- **Why has the bevelled edge chisel got sloping sides?**

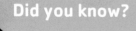
Tenon saw

Panel/tenon/dovetail saw

The three saws you will use most are the panel, tenon and dovetail saw. The largest of these three saws is the panel saw, which is used for cutting larger section timbers; it consists of a steel blade with teeth along its edge and a handle. Next in size is the tenon saw, which has a medium-width blade; this makes is steadier than the dovetail saw, so it can be used for cutting larger joints accurately. A dovetail saw is much smaller and has a thinner blade for cutting more accurate dovetails. You use each of these saws in the same way, drawing the saw back and forth with a pushing and pulling action, to cut through the timber fibres.

Planes

A jack plane is used for removing larger shavings of timber bringing its surface down to a smooth finish. A smoothing plane is used to smooth areas to an acceptable finish.

Activity: The block plane

A block plane is another type of plane. Find the answers to the following questions.

- **How big is the block plane?**
- **What is it used for?**

Provide a sketch of a block plane, labelling any parts.

Wheel brace

A wheel brace is used to hold a drill bit or auger bit and to turn it, drilling a hole through the timber.

Bradawl

A bradawl consists of a handle with a steel pit. This is used to start a hole in timber, for a drill bit to gain purchase or to accept a screw. The hole enables you to position the bit or screw accurately.

Hand screwdrivers

As its name suggests, a screwdriver is used to drive home a screw that has a thread. You do this clockwise, by fitting the screwdriver head into the head of the screw.

Activity: Types of screwdriver

There are two main types of screwdriver in use today. Find out what they are, including their technical names.

Bench holdfast

This special cantilever tool operates by screwing a bolt against a cam, which tightens down and fixes the timber securely to the bench so it will not slip as you work on it.

Assessment activity 8.1 · BTEC · P1

Imagine you have just started a joinery apprenticeship and have to purchase some hand tools. Using a local tool supplier's catalogue, identify the hand tools that are used to perform typical carpentry and joinery tasks. Photocopy these pages and create a collage of images and captions to show what you would buy and why. P1

Grading tip

To obtain P1 you will need to name the hand tools required for a given joinery or carpentry situation. This can be accomplished as a simple list.

PLTS

Reflective learner think about how you might use the tools in your practical work

Effective participator work out how to distinguish between similar tools

Functional skills

ICT use a computer to draw up your list of tools

1.2 Materials

Renewable softwoods

Wherever possible, you should use softwood timbers from managed forests in preference to those which are not, or hardwoods. Renewable softwoods are normally grown in colder climates, which produce wood with a straight, even grain because they grow more slowly.

Nails

Nails are manufactured from steel and are used to fix timber together. They are a rough fixing type and you use a hammer to drive them home. Nails come in a variety of types from heavy duty round-heads to lost-head oval nails.

Panel pins

Panel pins are small nails used to fix panels, as their name suggests! They are small enough to drive home with a panel pin hammer, which is a smaller version of a normal hammer.

Did you know?

Even off-cuts from using softwood can useful, as they can be turned into further timber products: for example, by chipping and turning into chipboard sheets

Woodscrews

Woodscrews are used to fix detailed joinery items. The woodscrew has a screw thread built into its structure. The thread and length of the screw come in varying sizes, and are given different numbers: for example, a no. 6 screw is smaller than a no. 10 screw. The larger numbers are used to fix heavier and larger timber sections. They can be manufactured in various metals (for example, zinc) and coated to protect them from rust.

Polyvinyl acetate glue

This white glue, which you might know by its shorter name, PVA, is used for reinforcing a timber joint. You must leave it to dry overnight, to form a hard bond between the timber fibres. Wipe up any excess glue using a damp cloth. Take care to prevent glue coming into contact with your skin.

Abrasive paper

This is paper to which a layer of glue is applied, then various grades of sand are sprinkled evenly over the surface and allowed to dry. It is used to sand down the surface of timber, making it smooth. Abrasive paper is available in many different grades. A coarse grade has much larger particles bonded to the paper and is used for rough sanding down of timber products; finer finishing grades are used to produce a quality surface.

Assessment activity 8.2 **BTEC** P3

You are part of a team taking on a range of carpentry and joinery tasks for a construction project. You have been given the job of sorting out which materials are needed. The foreperson gives you this list of tasks:

1 fastening a plywood back to a shelf unit
2 connecting one timber roof joist to another
3 fixing a door to a frame using hinges
4 providing additional strength to a joint
5 providing an intermediate first floor.

What materials will your team need for each of these? P3

Grading tip

To achieve P3 you will need to name the materials required to accomplish each carpentry and joinery task. A simple descriptive list is all that is required.

2. Health, safety and welfare for carpentry and joinery

2.1 Health, safety and welfare issues

Maintaining a clean and tidy work space

An untidy site or workbench can cause several issues, including:

- potential fuel for a fire if working near hot trades
- a trip hazard
- potential falling objects, if at height
- cuts from sharp tools.

After each operation, or at the end of the working day, you must make time to clean up the workspace so that:

- all waste materials are removed to a bin
- tools are placed back into a tool box
- any excess materials are taken back to the store.

Brushes, vacuum cleaners, bins and shovels should be provided for this purpose.

If you continue this good housekeeping practice, you will reduce the risk of an accident and make the workplace a better environment to work in.

Did you know?

Slips and trips are regularly the biggest single cause of reported injuries in the construction industry, causing over a 1000 major injuries each year.

2.2 Hazards

Identifying hazards

To stop an accident occurring, you must be able to identify the hazard that may cause the accident. You will remember that a hazard is something that has the potential to cause harm. In Unit 2 (page 28), you saw how hazards are often classified as:

- physical
- environmental
- chemical
- biological
- psychosocial.

Some of the chief hazards you will come across in carpentry and joinery will be physical, chemical or biological.

⚠ **Cross reference**

See Unit 2 pages 28–40 for more about risk assessment and controlling hazards.

Key term

Tetanus – a rare but often fatal disease that affects the central nervous system, caused when bacteria enter the body through a wound or cut.

Did you know?

Tetanus is often associated with rust.

- Physical hazards are identified through experience of working with materials and in working environments. Example: a loose handrail on an access platform.

- Chemical hazards come from the use of chemicals. Example: PVA glue over time causing dermatitis to the skin on your hands.

- Biological hazards are often ones that you cannot see. Example: **tetanus** during demolition operations.

Once you have identified a hazard, you need to evaluate its potential risk. This means looking at the hazard and deciding who might be harmed, and what could be the result of this. Hazards need to be controlled so they do not cause a high risk. Controlling hazards is not the same as using PPE, which should always be your last resort.

Slips, trips and falls

As you have seen, slips, trips and falls are major causes of accidents during construction activities. Falling from height is the major cause of fatalities on sites. It is the simplest of accidents, but can have a major effect on a person's working life.

Case study: One small step for John

John was a joiner working for Simon Construction on a factory extension. The site supervisor asked him to set out a row of blocks for the bricklayers to start working on in the morning. John finished this work, cleared his drawings and tape measures away and walked over to the exit. As he stepped over the row of blocks he had laid out, he caught his foot and fell to the ground, breaking two bones in his arm.

John was off work for eight months, and had to have major surgery before he had the strength to use the arm again. This was all due to tripping over a block that was 225 mm high.

- **What was the root cause of the accident?**
- **How could this accident have been prevented?**

Cuts and injuries caused by sharp tools and instruments

You hold most sharp tools in your hands, so it is your hands that are most likely to get cuts. Tools must be kept sharp and in a good working condition, as blunt, broken tools will cause accidents. If a tool is blunt, the tool will not grip properly and you may have to exert more pressure to cut the timber, which may result in slipping and a cut. Avoid carrying sharp tools, or do so with great care, as they can be a danger to others if you drop them.

Musculoskeletal injuries resulting from lifting and moving heavy loads

Lifting heavy objects can cause back pain as a result of damage to the soft tissues of the spine. This is the main **musculoskeletal** damage that can happen if you do not lift objects correctly.

You should always use kinetic lifting techniques. Keep your back straight and your knees bent and lift with the legs, making sure that you keep the object you are lifting close into your body.

Always assess the weight of the object that you are going to lift for risk and, wherever possible, use mechanical rather than manual means. Make sure that timber is delivered using a crane offload or a forklift, as this saves some of the manual strain from lifting.

2.3 PPE

This unit focuses on the main types of PPE associated with carpentry and joinery.

Safety boots

Safety boots contain a steel-reinforced toecap and sole plate. This prevents crush injuries to the toes and puncture injuries to the base of the foot. This essential piece of PPE should be worn at all times when you are working on a construction site, or as directed by a supervisor.

Hand protection

You should protect your hands by using the correct type of glove for the operation you are undertaking. For example, if you are working with glue, you will need a glove that will stop the glue bonding to your hand, but will not be affected by the glue itself. Gloves are manufactured in many varieties for different tasks. Heavy-duty rigger's gloves are the main type used on construction sites, as they help protect the hand and have a heavier material on the palm to resist splinters and abrasion to the skin. Rubber gloves are available for use with liquids.

Goggles

Goggles are used to provide full eye protection as they cover the eyes from the front and the side with an enclosed plastic mask. Goggles stop flying objects from entering the eyes and causing an injury. You would need goggles for tasks such as using a circular saw.

Anything else?

You should always wear a safety helmet when working on site to prevent injury to your head from falling objects, low ceiling access and any other obstructions.

Key term

Musculoskeletal – to do with the human frame and muscles that functions to give movement.

Did you know?

The charity BackCare estimates the annual cost of back pain to the NHS, business and economy is £5 billion.

Cross reference

Look at Unit 2 for more information about PPE.

Key term

Justify – give a good reason for.

PLTS

Independent enquirer investigate the safety aspects of the two different tasks

Functional skills

English write a clear, detailed explanation of your choices

BTEC ## Assessment activity 8.3 P5 P6 M2

You are doing some carpentry and joinery tasks on a small building project, and your boss has asked you to do these two jobs:

- cutting a dovetail joint
- fixing two roof members together.

1 Identify the PPE and safe working practices used to perform the two tasks. **P5**

2 Explain why you selected the PPE and the safe working practices for the above two tasks. **P6**

3 **Justify** your choice of PPE and safe working practices to minimise health, safety and welfare risks for the two tasks. **M2**

Grading tips

1 To achieve **P5** you will need to name the different items of PPE required to perform the carpentry and joinery task safely. You will also need to give a simple method statement on the safe working procedure required for the task.

2 To achieve **P6** you will need to give a deeper written explanation, saying why you would choose each item of PPE and safe working practice you named before.

3 To achieve **M2** you need to give more detailed reasons for your choice of safe working practice and selection of PPE for the task. This might involve mentioning why you chose this item rather than another item, or describing relevant factors in the particular scenario, such as it involving working at height or with dangerous chemicals.

3. Safe working practices to mark out and form joints for a timber frame to a given specification

3.1 Marking out

A setting out rod is a full-size drawing, usually marked out on hardboard, from which you take all the measurements required to manufacture a joinery product. It is essentially a guide, where all the dimensions have been worked out beforehand so that any problems can be identified and corrected in advance.

Your tutor will have provided you with the details of the joints that you have to create; you will now need to prepare a setting out rod to work from.

Production and use of setting out rods

Vertical and horizontal sections of simple frames

When you have practised the construction of the woodwork joints, you will move on to preparing a simple timber frame, where each corner or junction will contain one of the joints that you have explored.

To do this, you will need to prepare vertical and **cross-section** setting out rods on a flat surface, so that you can evaluate how accurate your work is.

> ### Key term
>
> **Cross-section** – what the inside would look like if you cut through it.

Using setting rods when producing specified timber products

Any timber product has to contain a joint. This can be as simple as a butt joint with a nail to hold it in place. This joint will require quite accurate setting out. Using a setting out rod or drawing makes this procedure much simpler as you have a template to check against each time you make a cut and as you assemble the joint. Setting out rods are an invaluable tool for the joiner and carpenter.

Once you have developed the necessary skills in marking out, cutting and completing some of the basic joints you can move on to producing some timber products by way of progression. These may be scaled for the reasons of economy and handling of the sample timber product that you construct.

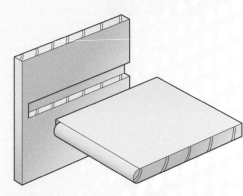

A housing joint is a slot that accepts the other timber to form the joint

3.2 Joints

The following drawings illustrate the typical joints that you may be constructing in the practical sessions, and show in detail how they are constructed.

Housing

A housing joint 'houses' the other piece of joiner timber into a groove or slot. This joint is often used for building shelving units in furniture.

Through/corner halving

This joint is a form of overlap, with half of each section taken away so they overlap and form a joint. Corner halving can be secured using pins, dowels and glue.

Tee halving

Tee halving forms the shape of a 'T' when it is completed. It follows the same principle as corner halving.

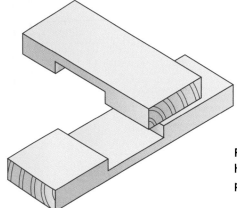

For a halving joint, like this tee halving joint, slots are cut in both pieces of timber so they fit together

Through/corner bridle

This joint is very similar to a mortise and tenon but the mortise is cut the full width of the tenon. As you can see, it is used to join frames at the corners.

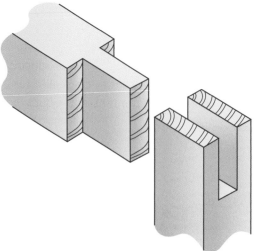

With a corner bridle, a slot is cut to accept a tenon

Through/haunched mortise and tenon

A through or haunched mortise and tenon is where the tenon passes right through the mortise and flush to the outside. This is often then wedged and pinned.

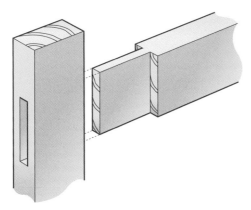

With a mortise and tenon, the tenon goes right through the mortise

Dovetail halving

This is a complicated joint commonly found in high-quality furniture, such as drawers. It is a halving joint with a cut shaped like a dove's tail within it, which resists lateral forces to pull it away from the other jointed timber.

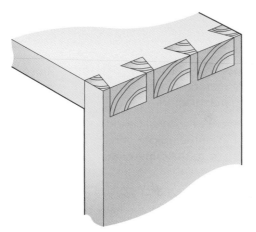

Look at the detail in this dovetail joint

3.3 Timber frame

The final assessment piece for this unit will be the production of specified simple frame made from prepared timber sections, including a range of the joints that you have explored to a given specification.

Your tutor will give you details of the frame that is to be manufactured, the sizes, tolerances required and what joints it has to contain. You will be required to plan, select, manufacture and assemble this frame to the specification.

BTEC Assessment activity 8.4 P2 P4 P7 P8 P9 M1 M3 D1

You have been asked to produce a small frame test piece that is 300 mm by 200 mm in size. It will incorporate a through tenon joint, a corner halving joint and a haunched mortise and tenon.

1 Select the hand tools you would need to undertake each of these three joints. **P2**

2 Select the materials required to perform this task. **P4**

3 Produce a setting out rod with vertical and horizontal section for the frame. **P7**

4 Set out and cut the three joints in the timber, producing the frame to specification. **P8 P9**

5 Justify your use of hand tools and selected materials to minimise health, safety and welfare risks. **M1**

6 Produce finished work with all joints within 3 mm tolerance and square. **M3**

7 Produce finished work with all joints within 1 mm tolerance and square. **D1**

Grading tips

1 To achieve **P2** you must select the correct hand tools for the task and demonstrate that you have done this, either by a written brief or by answering questions from your tutor, witnessed in an observation record.

2 To achieve **P4** you will need to show that you can select the right materials for the product that you have been asked to manufacture.

3 To achieve **P7** you will need to produce a setting out rod drawing, on either paper or hardboard, making sure that it is accurate to the dimensional information you have been given.

4 To achieve **P8** and **P9** you will be assessed on the practical outcome that you have been asked to produce. You will need to show the correct setting-out procedure for a joint and then cut it accurately, combining it within the timber frame sample.

5 To achieve **M1** you will need to expand on the first two criteria, giving reasons why you have selected these hand tools and materials, relating your answers to how this supports health and safety.

6 To achieve **M3** you must accurately cut the joints within 3 mm of the setting out rod you have produced. This is all about quality and accuracy, so take special care.

7 To achieve **D1** you will need to be even more accurate and pay close attention to detail, so that your joints meet an even tighter quality standard, with accuracy down to 1 mm.

WorkSpace Nigel Lynch
Joiner's mate

Nigel has been working as a joiner's mate for the past five years. He assists the site carpenter and joiner with heavy lifting or complicated assembly, where one person could not accomplish the task safely on their own.

Nigel's job involves lifting floor joists into position, fitting roof trusses and their bracing, handling flooring sheets and assembling stud partitions. This is hard manual work on site, in all weathers.

Nigel has been discussing his options with his line manager at the company he works for. His manager has suggested that he considers taking the first steps on the ladder to a professional qualification in site joinery and carpentry. This is the National Vocational Qualification route, or NVQ for short.

Nigel would start on the NVQ Level 2, attending a local training organisation one day per week to learn the theory and put it into practice in a workshop. For the rest of the week he would be working, using this time to gather evidence of other work he has done on site and building up a portfolio of evidence.

Nigel is slightly concerned about the amount of writing he will have to do. After discussing his concerns with his training provider, he will also complete a literacy qualification to help him with this area of his skills. Nigel now feels reassured about the future and has agreed to start the training at the next opportunity.

Because the NVQ is a registered and recognised course with Construction Skills, Nigel's employer can become involved as a work-based recorder. Now Nigel has a person at work who can check his evidence and sign the witness statements and photographs that Nigel collects as he completes certain tasks.

Nigel feels positive about the future; he has several opportunities open to him to advance within the company and better his current position.

Think about it!

1 What would you do in Nigel's position?
2 How would you find out about training opportunities?
3 What do you think the future holds for Nigel?

Just checking

1 Why is a combination square so useful?
2 What are the hazards involved in using hand tools?
3 Where does a claw hammer get its name?
4 There are two different types of cross-section used in chisels. What are they?
5 What different types of screws are there?
6 What are the main categories of hazard in carpentry and joinery?
7 Explain to a partner what PPE is and how it is used, so they can check you are right.
8 Sketch a halving joint.
9 What is a dovetail joint used for?
10 Explain how a setting out rod is used.

edexcel

Assignment tips

- Practice makes perfect when working with timber, so learn from the mistakes that you make and improve at the next attempt.

- Remember that you cannot stick timber back together once it has been cut, so be sure of your measurements before you make a cut.

- Make sure that the gaps in the joints of your finished wood products are not larger than the tolerance limits given for the piece of work.

9 Performing joinery operations

Joinery is a workshop skill involving the cutting and forming of joints in timber using hand and machine tools on a workbench. Joinery, as the name suggests, is the skill of forming joints in timber to connect them together. A joiner tends to be based in a workshop, whereas being a carpenter is a site-based role, often fixing items that a joiner may have manufactured in the workshop.

Joinery therefore requires some detailed skills in working with timber. These skills include being able to:

- accurately mark out work
- cut timber accurately
- keep work clean and free from imperfections
- understand the complexities of the timber structure.

Learning these skills takes time, trial and error, and practice in developing the manual dexterity to perform the joinery operations involved in producing quality timber products.

Learning outcomes

After completing this unit, you should:

1 know the hand tools and materials commonly used to perform joinery tasks

2 understand the important health, safety and welfare issues associated with joinery tasks

3 be able to apply safe working practices to mark out and form joints for a timber product.

Assessment and grading criteria

This table shows you what you must do in order to achieve a pass, merit or distinction grade, and where you can find activities in this book to help you.

To achieve a **pass** grade the evidence must show that you are able to:	To achieve a **merit** grade the evidence must show that, in addition to the pass criteria, you are able to:	To achieve a **distinction** grade the evidence must show that, in addition to the pass and merit criteria, you are able to:
P1 identify the hand tools used to perform joinery tasks **See Assessment activity 9.1, page 87**	**M1** justify the safe use of hand tools and materials to minimise health, safety and welfare risks **See Assessment activity 9.4, page 94**	
P2 select the hand tools required to perform given joinery tasks **See Assessment activity 9.4, page 94**		
P3 identify the materials used to perform joinery tasks **See Assessment activity 9.2, page 88**		
P4 select the materials required to perform given joinery tasks **See Assessment activity 9.4, page 94**		
P5 identify the PPE and safe working practices used to perform joinery tasks **See Assessment activity 9.3, page 91**	**M2** justify the safe use of appropriate PPE and working practices to minimise health, safety and welfare risks **See Assessment activity 9.3, page 91**	**D1** produce finished work safely with all joints within 1mm tolerance and square **See Assessment activity 9.4, page 94**
P6 explain the selection of the PPE and safe working practices to be used in given joinery tasks **See Assessment activity 9.3, page 91**		
P7 produce setting out rods and use them to mark out timber **See Assessment activity 9.4, page 94**	**M3** produce finished work safely with all joints within 3mm tolerance **See Assessment activity 9.4, page 94**	
P8 set out and cut joints in timber **See Assessment activity 9.4, page 94**		
P9 use a range of joints to produce a panel door or a casement window to a given specification **See Assessment activity 9.4, page 94**		

How you will be assessed

This unit will be assessed by an internal assignment that will be designed and marked by the staff at your centre. Your assessment could be in the form of:

- presentations
- case studies
- practical tasks
- written assignments.

Gemma, 15 years old

I knew a little about joinery already because I took Unit 8 Exploring carpentry and joinery, and for that we'd looked at some basic joints and I really enjoyed producing these in a small frame. I took this unit because it will extend my knowledge and understanding of joinery and take my skills to the next level in terms of detail and complexity.

It's been great to have a look at some of the more complex joints that are used to hold timber sections together. I have learned that you must take great care in the marking out of the joint correctly, marking the timber that has to be removed and cutting the joint accurately.

This unit has also made me aware of the accuracies that are required in joinery in order to produce a quality timber. Quality is very important if you are going to produce a product a client would want to buy, and it's satisfying when you get it right. You have to take care with the finish too, as products might require a protective layer of sealer, like varnish.

Over to you!

- Which do you think might suit you best: carpentry or joinery?
- What safety issues do you think this unit might cover?
- What skills do you think you might need to be a good joiner?

1. Hand tools and materials commonly used in joinery tasks

Build up

That's quality

Joinery is all about producing a quality product for the client. Care has to be taken about the way a finished product will look when it is installed: for example, will the finished joints need to be seen? Will they require a dry joint or do you need to use glue?

- **What other things would you need to consider as part of making a good quality product?**

- **How can you finish the work so it is clean and presentable?**

⚠ Cross reference

See Unit 8 for more information.

1.1 Hand tools

As a joiner, you will need to know what these tools look like and how they work:

- tri-square
- combination gauge
- marking/mortise gauge
- sliding bevel
- wooden mallet
- claw hammer
- mortise/bevel-edged chisel
- jack plane
- hand screwdrivers.

Before reading on, look at the descriptions, details and some photos of these tools in Unit 8, pages 67–68, then read the additional information in this unit about how these tools are used in joinery.

Tri-square

The tri-square is used to produce a 90° angle across a piece of timber. You should take great care when using it to make sure it does not get damaged; if it does, it will not give you a true 90° angle against the face of the timber, which will lead to poor quality workmanship.

Combination gauge

A combination gauge has two uses. As it has one pin on one side and two pins on the other, it can be used for marking out tenons using the one-pin side, and mortises using the two-pin side.

Tri-square

Combination square

This is a very versatile tool. It can be used as a simple tri-square, but it can also measure 45° angles and has a built-in spirit level. It can be used on site as well in carpentry work.

Marking/mortise gauge

These two types of gauge are used to mark out a line parallel with a face when dragged along. This gauge is especially useful for marking out mortise and tenon joints – for the tenon on one face and the mortise on another – so that they fit accurately.

Sliding bevel

A sliding bevel can be used to transfer angles and mark out any angle required for a joinery product.

Wooden mallet

You use a wooden mallet with a chisel to cut into timber.

Claw hammer

A claw hammer can be used to extract nails from timber.

Mortise/bevel-edged chisel

Mortise and bevel-edged chisels have different cross-sections for different uses. A mortise chisel is used to cut out the slot in a joinery product that accepts the tenon. It enables the bottom of the slot to be kept clean and straight, so the tenon fits all the way home. A bevel-edged chisel has sloping sides, so it enables you to cut right down to the line on a dovetail. You could not do this with a straight-edged chisel.

Chisels are available in different width sizes for different operations: for example, fitting a mortise lock to a door or cutting out for a dovetail joint.

Did you know?

Quality marking gauges are made out of a timber called rosewood.

Look at the cross-section of this mortise chisel.

Activity: Chisels

A mortise chisel often has a ferrule on the end of the handle, to prevent the handle splitting (you tend to use a wooden mallet with this type of chisel). Chisels are sharpened with a stone that grinds the blade, using a lubricant, to a fine point.

- **The bevel-edged chisel is available in a range of sizes. Which are the most common sizes in use?**
- **A chisel handle is often made from boxwood. Why?**
- **How are both types of chisel sharpened?**

Panel/tenon/dovetail saw

Dovetail saws are fine saws with many teeth per millimetre. This enables you to cut a joint very accurately, which is especially useful for dovetails, which need many accurate cuts. If you tried this with a panel saw, which has fewer teeth per millimetre and is much larger, it would be hard to control and your work would not be accurate. For panel and tenon saws, see the descriptions and photo in Unit 8, page 68.

Block plane

A block plane fits into the palm of your hand, and is used for fine finishing of timber by removing the smallest amount at a time. It is often used for the small, out-of-reach places that a normal plane cannot get to.

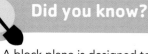

Did you know?

A block plane is designed to cut across the end grain of timber.

Activity: Using a plane

A plane has to be sharp in order to work efficiently. Find the answers to the following questions.

- **How do you remove the blade for sharpening?**
- **How do you control the depth of the cut?**
- **How do you keep the blade parallel with the plane base?**

Smoothing plane

Smoothing and jack planes

The smoothing plane is much smaller than the jack plane. It is used to smooth small areas to an acceptable finish over a large width, whereas a jack plane is used for removing large areas of timber.

Wheel brace and bradawl

As a joiner, you will also need to know about the wheel brace and the bradawl. For more information about these tools look at Unit 8, page 68.

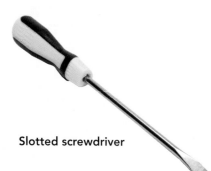

Bradawl

Hand screwdrivers

A screwdriver is used to drive home a screw that has a thread. There are two main types of screwdriver, to fit the two types of screw: Phillips (or crosshead), which has a cross-shape section, and standard (also known as straight or slotted), which has a simple slot-shaped section. Shorter screwdrivers are better for working in tight spaces, while longer ones give better turning power.

Slotted screwdriver

Sash clamp

A sash clamp is used to assist with glued joints on long or wide frames, holding together large pieces of manufactured joinery that a G-clamp (see below) is too small for. For example, with a sash clamp, the full width of a door can be clamped while the joint dries. The long bar or 'sash' has holes in it and a clamp head that slides along the bar, and can be stopped with a pin.

G-clamp

A G-clamp gets its name from its frame and screw, which looks like a letter 'G'. Available in various sizes, it is used to secure timber to a bench or to fasten and hold pieces together while they are glued.

G-clamp

Bench hook

Bench hook

You use a bench hook to hold timber steady while you cut it with a tenon saw. The bench hook is often placed in a vice to hold it, while you push the timber you are cutting against the top. You can then make a cut across the timber without slipping and causing an accident.

Spirit level

A spirit level has a curved glass or plastic tube filled with liquid, in which a bubble of air is suspended. As air is lighter than the liquid, the bubble will always float to the top of the centre of the curve. When the tube is built into a frame and set correctly, you have a completed spirit level. You can use it to check surfaces for level, by making sure the bubble sits in the centre of the tube. Spirit levels are available in a range of sizes to suit the use required.

Assessment activity 9.1

BTEC **P1**

You have just started a bench joinery apprenticeship and have to identify some hand tools associated with joinery work. Using a web-based tool supplier's catalogue, identify the hand tools that are used to perform typical specific joinery tasks, such as cutting a dovetail joint.

Cut and paste these into a document with the correct name against each. **P1**

Grading tip

To achieve **P1** you must name the hand tools needed, with a small explanation against each. These need to be correct for the joinery task that you are undertaking.

PLTS

Creative thinker think about how you might use the tools in your practical work

Self-manager organise your work carefully and efficiently

Functional skills

ICT use a word-processing package to create your document

1.2 Materials

You need to know about the main materials involved in joinery: renewable softwoods, nails, panel pins, woodscrews, polyvinyl acetate glue and abrasive paper. Before reading on, look back at Unit 8, pages 69–70, for more information about these. This section gives you some additional information relating specifically to joinery.

Panel pins

Panel pins can be made from different materials. Where the pin head is to be seen on a fitting, they are often manufactured from decorative brass. A panel pin can be easily punched below the surface as it has a small head. The hole can then be filled before a decorative finish is added.

Polyvinyl acetate glue (PVA)

This type of glue dries better when under pressure, so you should use a G-clamp or a sash clamp to obtain the maximiun strength for a joint. Excess PVA glue must be cleaned off a timber product using a damp cloth before it has set; if you don't do this, it will leave a stain on the face of the timber.

Cross reference

See Unit 8 for more information.

Did you know?

Before PVA, the glues carpenters and joiners used were animal glues, made from boiling down animal bones and carcasses.

PLTS

Creative thinker suggest alternatives to the usual materials

Effective participator suggest ways to break down each task

Functional skills

ICT use the computer to create a table of materials

BTEC **Assessment activity 9.2** **P3**

Working on a house extension, you have been asked to find suppliers for the materials needed for a list of small joinery jobs. Look at the list below, and identify the materials you would need for each piece of work.

1 Making a replacement drawer
2 Manufacturing a timber window frame sash
3 Making a panelled door
4 Manufacturing a door lining **P3**

Grading tip

To achieve **P3** all you need to do is to list the correct materials for the required task.

2. Health, safety and welfare for joinery tasks

2.1 Health, safety and welfare issues

Maintaining a clean and tidy work space

An untidy workshop or workbench can cause several issues, including these:

- shavings acting as fuel for a fire if you are working near hot areas

- trip hazards from waste materials on floor

- splinters to hands from unplaned timber

- cuts from sharp tools not stored correctly.

From reading Unit 2, you will already know how important it is to make time to clean up the workspace after each operation, or at the end of the working day. To remind you, you will need to make sure that:

- all waste materials are removed to a bin

- tools are placed back into a tool box

- any excess materials are taken back to the store.

If you continue this good housekeeping practice, you will reduce the risk of an accident and make the workplace a better environment to work in.

2.2 Hazards

Identifying hazards

You will remember from Unit 2 that a hazard is something that has the potential to cause harm.

Some of the chief hazards you may need to identify in joinery will be physical, chemical or biological.

- Physical hazards are identified through experience of working with materials and in working environments. Example: not stacking boards correctly, so that they fall.

- Chemical hazards come from the use of chemicals. Example: a wood primer painted onto the back of skirting boards before fixing.

- Biological hazards are often ones that you cannot see. Example: spores from wood that you can inhale.

As mentioned in Unit 8, once you have identified a hazard, you need to evaluate its potential risk, looking at the hazard and deciding who might be harmed, and what could be the result. Then you can try to find ways to control the hazard, before thinking about using PPE, which should be your last resort.

Did you know?

Dust from woodworking machinery can cause asthma, as it is an irritant to the lungs.

Cross reference

See Unit 2 pages 31–40 for more about **risk assessment** and **controlling hazards**.

Slips, trips and falls

These can easily occur in any workshop where materials are processed into a finished product. This can lead to trip hazards from loose materials left lying around, or where the workshop has not been cleaned properly.

Case study: Rick's rights

Rick was a bench joiner in a local joinery company. He had worked there for several years without incident. As they were always busy, the workshop was always in a mess with shavings, waste timber and other sheet materials just left lying around until someone had enough time to tidy up.

One day the Health and Safety Inspector called and undertook an inspection. He considered the joinery shop so dangerous that he closed it down, until such time as it was properly cleaned up and a safe system of work had been implemented. Rick lost a full week's wage as a result.

- **Has the Health and Safety Inspector the right to close down the workshop?**
- **Would Rick get his wages back?**
- **How could this be prevented from happening again?**

Cuts and injuries caused by sharp tools and instruments

You hold most sharp tools in your hands, so it is your hands that are most likely to get cuts. It is difficult to wear gloves when you are working timber, and sometimes wearing gloves can put you in greater danger. For example, if you are wearing gloves while using a battery drill, you cannot hold the screw to start it and you may slip. You need training and experience to use these tools correctly.

Musculoskeletal injuries resulting from lifting and moving heavy loads

Musculoskeletal damage from incorrect lifting can easily happen in a joinery shop. Joiners have heavy and awkward sheet materials to lift: for example, a 2.4 m × 1.2 m sheet of 22 mm plywood is very heavy to lift on your own; to lift it correctly, you would need two people. Larger joists and heavy timber sections would also need two people to lift.

As a joiner, you must know the correct way to lift, to prevent musculoskeletal injury. Look back at Unit 8 page 73 for more information.

Did you know?

Four in five adults experience back pain at some point in their life.

Cross reference

Look at Unit 2 page 34 and Unit 8 page 73 to remind yourself of the PPE that all carpenters and joiners need to be aware of.

2.3 PPE

You will have learned about PPE in Unit 2 and looked at the general sorts of PPE used in carpentry and joinery in Unit 8. Here are some additional notes on the particular PPE involved in joinery tasks.

Safety boots

Safety boots, which have a steel-reinforced toecap and sole plate, should be worn at all times when you are working in a workshop, to prevent injuries from dropping materials on your feet.

Hand protection

Using the correct type of gloves for the task is vital. If you are working with rough-sawn timber, you may need a heavy-duty glove.

Goggles

Goggles are important for preventing injury during many joinery tasks, such as using a circular saw or a planer. However, in a workshop environment, safety glasses often give your eyes adequate protection.

Anything else?

If you are standing next to machinery or using power tools, you may need to protect your hearing using ear protectors, to reduce the level of noise that gets through.

PLTS

Independent enquirer judge the value of different PPE for reducing risks

Functional skills

ICT use the internet to research safe working practices

BTEC **Assessment activity 9.3**

A local store has asked your joinery shopfitting firm to undertake some work in their premises. They require some additional shelves fitting to a wall and the front timber door glass replacing, as it is cracked.

1 Identify the PPE and safe working practices you should use for these tasks. **P5**

2 Explain why you selected the PPE and the safe working practices for the two tasks. **P6**

3 Justify your choice of PPE and safe working practices to minimise health, safety and welfare risks for the two tasks. **M2**

Grading tips

1 **P5** has two sections to it. To achieve it, first you need to correctly name the PPE required for the two activities that are going to be undertaken; then you need to identify the safe working practices that you will use during the tasks in the shop.

2 To achieve **P6** you will need to explain each of your choices of PPE and safe working practices, with some description against each showing why you chose it.

3 To achieve **M2** you need to give reasons why you chose these and explain just how they minimise the risks involved.

Cross reference

Look at Unit 8 for more information.

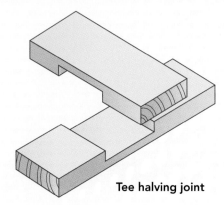

Tee halving joint

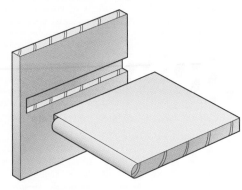

Housing joint

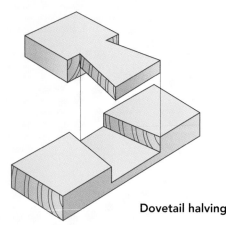

Dovetail halving

Did you know?

Furniture in Egyptian tombs was held together using dovetail joints.

3. Safe working practices to mark out and form joints for a timber product

To carry out the assessment for this unit, and to achieve the grading criteria, you will need to know about marking out, producing and using setting out rods. Before reading on, look at Unit 8 page 75 for the information you need on these areas.

3.1 Joints

You have already read about the different types of joint in Unit 8 (pages XX-XX). Take some time to re-read this section now, then read on to look at the joints particularly related to this unit's work.

Corner/tee halving

Corner halving is a form of overlap, with half of each section taken away so they overlap and form a joint. Corner halving can be secured using pins, dowels and glue. Tee halving follows the same principle as corner halving, but forms the shape of a 'T' when it is completed.

Housing

A housing joint 'houses' the other piece of timber into a groove or slot. This joint is often used for building shelf units in furniture.

Dovetail halving

This is a halving joint with a cut shaped like a dove's tail within it, which resists lateral forces to pull it away from the other jointed timber.

Mortise and tenon (wedged)

You found out about mortise and tenon joints in Unit 8, page 77. A wedged mortise and tenon joint is a very strong joint. Small wedges are driven home, using glue to expand the tenon, and tightened to make the whole joint secure. After the glue has dried, you clean up the joint by cutting off the ends of the wedges and sanding them down.

Double/twin mortise and tenons (wedged)

This type of joint has not just one mortise and tenon, but two. It acts just like a single joint, and is used for additional strength in door construction. These joints can also be wedged to make them even stronger.

3.2 Timber products

Panel door

A panel door, as the name suggests, contains panels. These are secured within frames that are fixed together using the joints that we have looked at and examined previously.

Casement window

A casement window is one that contains a window attached with hinges, so it has a frame with a window attached within it. This sash frame is attached to the casement window using hinges and ironmongery. A casement window is usually formed of softwood timber.

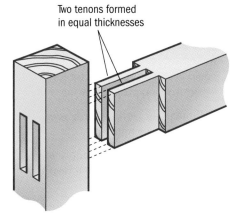

Two tenons formed in equal thicknesses

Double mortise and tenon joint

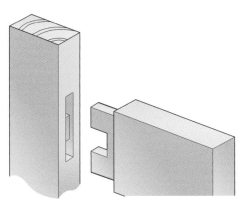

Twin mortise and tenon joint

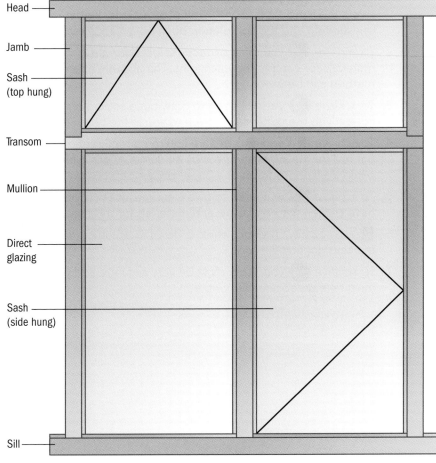

Head

Jamb

Sash (top hung)

Transom

Mullion

Direct glazing

Sash (side hung)

Sill

In a casement window, how is the sash secured to the frame?

BTEC **Assessment activity 9.4** P2 P4 P7 P8 P9 M1 M3 D1

You have been asked to produce a scaled small timber casement window using a range of joints, which will incorporate some of the joints that you have practised.

1 Select the hand tools you would need to undertake each of these three joints. **P2**

2 Select the materials required to perform this task. **P4**

3 Produce a setting out rod with vertical and horizontal section for the frame. **P7**

4 Set out and cut the three joints in the timber producing the window frame to specification. **P8** **P9**

5 Justify your use of hand tools and selected materials to minimise health, safety and welfare risks. **M1**

6 Produce finished work with all joints within 3 mm tolerance and square. **M3**

7 Produce finished work with all joints within 1mm tolerance and square. **D1**

Grading tips

1 To achieve **P2** you will need to show that you can select the hand tools for each joint correctly, by verbal questioning and the completion of an observation record.

2 To achieve **P4** you must correctly select the right materials for completing each joint. The above evidence method can be used.

3 To achieve **P7** you will need to draw out a setting out rod correctly to the dimensions given to you in the brief.

4 To achieve **P8** and **P9** require you to correctly set out the rod for the joints and provide evidence of this, which could be a photograph or a paper copy.

5 To achieve **M1** you need to give clear reasons why you have chosen the hand tools and materials for this task.

6 **M3** is a quality criterion, which is down to accuracy in marking out and cutting.

7 To achieve **D1** your work must be very accurate and you must take great care with detail.

WorkSpace Andy Barker

Joiner

Andy works for a joinery manufacturer in his home town. He has worked for this company for a number of years. They produce one-off manufactured joinery products for the shop fitting industry. These are items such as reception desks, shelving units, cupboards, changing rooms and sales desk units.

Andy really enjoys the work. It is varied and he has the job of coordinating all the dimensions of the timber products that they make. This often involves him travelling to different sites where the company's products are going to be installed, and taking specific and detailed site measurements to establish the correct size for the finished product, and whether it will fit through the access doorways provided!

Andy has had to develop some graphical skills so he can record these detailed measurements and show how the product should be constructed. He must do this in a clear and detailed way so that the joiners in the company's workshop understand the assembly drawings and produce a product to the correct specifications. Andy also needs to be good at mathematics as he has to work out the quantities of materials that need to be purchased for each project.

To strengthen his skills, Andy has decided to return to college and attend an evening course on computer-aided design (CAD). This is the technology architects and designers use to create the drawings that they send to the joinery estimator to price work. Andy makes his drawings by hand but realises that eventually electronic drawings will make their way into the production side of the business and he needs to fill this knowledge gap.

Andy was apprehensive about attending the course as it had been a long time since he attended formal classes, but is pleased to begin to understand the electronic programs and tools that can be used by business and industry to produce technical drawings.

Think about it!

1 Would you take the steps that Andy has?
2 What are the possible benefits of doing this?
3 What opportunities will this open up for Andy?

Just checking

1 Why is a combination marking gauge different from an ordinary gauge?
2 In what way is a smoothing plane different from a jack plane?
3 Where would you use a block plane?
4 What is a bench hook used for?
5 Name three workshop hazards.
6 How is a corner halving different from a tee halving?
7 What is the difference between a dovetail halving joint and a dovetail joint?
8 How does a wedge strengthen a joint?
9 What PPE do you require when working in a workshop?
10 What is a tolerance and how do you achieve it?

edexcel

Assignment tips

- Practice makes perfect when working with timber. Don't be disappointed at your first attempt, as you will learn from the mistakes that you made and improve at the next attempt.

- Remember to 'measure twice, cut once', as you cannot stick back the timber you have just removed.

- Take care with your work and try to observe the tolerances given, as this will improve your marks.

- Keep your work clean and free from unnecessary pencil marks.

- Always keep your work area tidy.

10 Performing carpentry operations

Carpentry tends to be a site-based skill. Carpenters will usually be working on the construction site itself, fixing components that are assembled on site: for example, constructing a roof out of timber components. The components that the carpenter fixes may have been manufactured by a joiner in a workshop.

Carpenters often build quite specific, purpose-made items, such as ships, roofs, floors or film sets. Their work is mainly concerned with the construction stage.

Carpentry requires some detailed skills in marking out timber ready to cut. For example, a carpenter would need to know how to cut out the correct angles on a traditional roof.

The skills a carpenter needs include being able to accurately mark out work and understand drawings, cut timber accurately and understand the complexities of the timber structure.

Carpenters usually train by serving an apprenticeship so, over several years, they can build up the site skills they need to be able to site-assemble components.

Learning outcomes

After completing this unit, you should:

1 know the hand tools and materials commonly used to perform carpentry tasks

2 understand the important health, safety and welfare issues associated with carpentry tasks

3 be able to apply safe working practices to perform carpentry tasks.

Assessment and grading criteria

This table shows you what you must do in order to achieve a pass, merit or distinction grade, and where you can find activities in this book to help you.

To achieve a **pass** grade the evidence must show that you are able to:	To achieve a **merit** grade the evidence must show that, in addition to the pass criteria, you are able to:	To achieve a **distinction** grade the evidence must show that, in addition to the pass and merit criteria, you are able to:
P1 identify the hand tools used to perform carpentry tasks **See Assessment activity 10.1, page 102**	**M1** justify the safe use of hand tools and materials to minimise health, safety and welfare risks **See Assessment activity 10.4, page 108**	
P2 select the hand tools required to perform given carpentry tasks **See Assessment activity 10.4, page 108**		
P3 identify the materials used to perform carpentry tasks **See Assessment activity 10.2, page 102**		
P4 select the materials required to perform given carpentry tasks **See Assessment activity 10.4, page 108**		
P5 identify the PPE and safe working practices used to perform carpentry tasks **See Assessment activity 10.3, page 106**	**M2** justify the safe use of appropriate PPE and working practices to minimise health, safety and welfare risks **See Assessment activity 10.3, page 106**	
P6 explain the selection of the PPE and safe working practices to be used in given carpentry tasks **See Assessment activity 10.3, page 106**		
P7 produce setting out rods and use them to mark out timber **See Assessment activity 10.4, page 108**	**M3** produce finished work safely with all joints within 3 mm tolerance and square **See Assessment activity 10.4, page 108**	**D1** produce finished work safely with all joints within 1 mm tolerance and square **See Assessment activity 10.4, page 108**
P8 set out and cut joints in timber **See Assessment activity 10.4, page 108**		
P9 use a range of joints to perform carpentry tasks to a given specification **See Assessment activity 10.4, page 108**		

How you will be assessed

This unit will be assessed by an internal assignment that will be designed and marked by the staff at your centre. Your assessment could be in the form of:

- presentations
- case studies
- practical tasks
- written assignments.

James, 15 years old

Carpentry and joinery? I didn't really know the difference until I did this unit, but now I can see that carpentry is its own thing, with its own skills. For this unit, we looked at some basic joints and I got to have a go at producing these in a small frame. I really felt like I had extended my knowledge and understanding and got to grips with some of the real, practical aspects of carpentry, which I can now use for work and at home.

The carpentry work involved us looking at hanging a door to ensure that it is level and vertical. It was much harder than it looked! We even fixed the lock and hinges to complete the full installation. It was good to be able to do the whole thing, from start to finish.

I hadn't really thought before about being able to work outside with carpentry, but now I realise you can, on things like roofs and floors. It's this aspect of the work that I would really enjoy the most, as I love being outdoors.

Over to you!

- How do you think carpentry will compare with joinery?
- What experience have you already had of carpentry?
- What do you think the most challenging aspects of carpentry might be?

1. Hand tools and materials commonly used in carpentry tasks

Build up

First in line

Carpentry is often about producing a foundation that other trades work off: for example, installing timber partitions for the electricians to wire through and the plasterers to board over.

- What other items would a carpenter install?
- What are the crafts that would follow, after the carpenter has finished?

⚠ **Cross reference**

Look at Unit 8, pages 66–69 and Unit 9, pages 84–87 for descriptions, details and some photos of these tools.

1.1 Hand tools

For carpentry, you will need to know about a range of tools including pencil, marking/mortise gauge, sliding bevel, wooden mallet, claw hammer, mortise/bevel-edged chisel, tenon/panel/dovetail saw, jack/smoothing/block plane and hand screwdrivers.

Many of the tools you will need for carpentry tasks are covered in Unit 8 and Unit 9. Look back at these sections before reading on, then read this additional information about the tools that are particularly relevant to carpentry.

Smoothing, block and jack planes

The plane is a key tool for carpentry tasks. A smoothing plane is used to smooth areas to an acceptable finish. A block plane is used to remove the smallest amount of timber at a time. A jack plane is used for removing larger shavings of timber bringing its surface down to a smooth finish.

Jack plane

Wheel brace

Wheel braces are used to hold a drill bit or auger bit and to turn it, drilling a hole through the timber. This is especially useful for removing timber when you are fitting a lock case, where you would drill top and bottom then drill out the remaining middle and chisel out the sides flat.

Bradawl

A bradawl is a handle with a steel pit. A bradawl is useful to prevent a drill bit or screw sliding away from the position you want it to start in.

Activity: The parts of a plane

Before starting, make sure that your tutor has given you permission to undertake this activity.

- Lay a plane down onto a flat workbench.
- Release the iron clamp and take it out of the plane mouth.
- Remove the plane blade and clamp – take great care as this is sharp.
- Undo the blade clamp and remove.
- Investigate how the blade is moved up and down to increase the depth of cut.
- Investigate how the plane blade is kept parallel with the sole of the plane.
- Reassemble the plane parts.

Mitre box

A mitre box is used to cut a 45° cut across the grain of one timber section, so that it can be joined to another: for example, at the corner of a moulded architrave.

Nail pincers

Nail pincers are used to remove nails from timber. To use them, you clamp the nail in the jaws and lever the nail out by roiling them across the curved section.

Nail punch

You use a nail punch to drive a nail below the surface of a timber, so that it cannot be seen, and the hole can be filled and sanded smooth. You fit the nail punch over the head of a nail and then strike it with a hammer to drive it below the surface, which you can then decorate. The end of the punch needs to be smaller than the nail.

Cutting a mitre on a moulding

Nail punch

BTEC Assessment activity 10.1 · P1

You have been asked to hang a door made by a joinery company, using brass hinges.

Identify the hand tools that you will need to take with you in performing this operation. **P1**

Grading tip

To achieve **P1** you will simply need to list the hand tools that you will require to undertake the hanging of the door.

1.2 Materials

You need to know about the main materials involved in carpentry: renewable softwoods, nails, panel pins, woodscrews, polyvinyl glue and abrasive paper.

BTEC Assessment activity 10.2 · P3

You have just been taken on as a carpenter's assistant on a small building site. Your boss has asked you to find out what materials you will need to perform the following carpentry tasks:

1 hanging a replacement door

2 fitting a window frame

3 laying floor joists

4 building a stud wall.

Using this book and any other sources, make a list of what would be required for each of these jobs. **P3**

Grading tip

To achieve **P3** you will need to identify the correct materials that will be required to accomplish each of the listed tasks.

2. Health, safety and welfare issues associated with carpentry tasks

2.1 Health, safety and welfare issues

Maintaining a clean and tidy workspace

A construction work area needs to be kept clean and tidy and free from loose materials, which can cause a trip hazard. A suitable skip should be provided for the safe disposal of waste materials. Off-cuts of timber should be separated out and put into timber skips for recycling into further timber products.

Many carpentry operations are undertaken at height: for example, constructing a new roof on a two-storey house. You need to take care to make sure the working platform is kept free of trip hazards that could cause a serious injury.

As with joinery tasks, good housekeeping practice will reduce the risk of accidents and makes the workplace a better environment to work in.

2.2 Hazards

Identifying hazards

Some of the chief hazards you may need to identify in carpentry will be physical, chemical or biological:

- physical hazards include noise from cutting machinery
- chemical hazards include splinters from chemically treated timber causing blood poisoning
- biological hazards include the resins from some softwood sap materials.

Slips, trips and falls

These can occur anywhere in a construction environment. There are many different trades and operatives working on a site at anyone time. Everyone must be diligent to the trip hazard and clean up their own work areas.

Did you know?

If enough dust builds up in the air from woodworking machinery, it can cause a dust fire, which is extremely dangerous.

Cross reference

Look at Unit 2, pages 71–73, for more information on hazards.

Case study: Craig's loft lift

Craig, a carpenter, suffered serious injury at work after a fall from height which could have ended his career.

He was carrying a 2.40 m by 1.20 m, 30 kg piece of chipboard on an extension ladder during a loft conversion, when he fell, causing a broken wrist, broken ankle and injuries to his face where the chipboard struck him after the fall.

Craig's employers had been given advice previously by the HSE, but had still failed in their health and safety duties to provide a safe working method for transporting materials from ground level up to the loft.

- **What is a safe weight to lift?**
- **How would you safely handle a sheet material?**
- **What would you do to prevent this occurring again?**

Cuts and injuries caused by sharp tools and instruments

Carpenters normally wear a tool belt to keep sharp objects in, so that they can be kept safely while they are working. Chisels should have their plastic caps fitted when transporting them, not only to prevent an injury but also to keep them sharp. Where possible, the use of marking knives should be avoided as these are a primary cause of cuts to the hands and arms.

Avoiding injury when lifting or carrying

Take time to plan the task. Remember 'WILT': Working environment (tripping hazards, steps, doors, weather conditions), Individual capacity (injuries, appropriate clothing and footwear), the Load (does it have sharp edges? is it stable? does it contain hazardous materials?) and the Task (will you have to kneel/ bend? how many times will you need to move the load?).

Reduce the horizontal distance between the load and the body, get the waist as close as possible to the load, adopt a stable position and get a secure hold. Ensuring feet are stable, lean upper torso over the load, and straighten slowly, without twisting. While moving the load, make sure you can see where you are going, and stop for a rest if you need to.

Did you know?

Over 40% of accidents reported to the HSE involve manual handling.

Can you identify the actions this person is taking to lift safely?

2.3 PPE

Safety boots

Safety boots prevent crush injuries to the toes and puncture injuries to the foot.

Hand protection

When you are working with rough-sawn timber, wear appropriate gloves to prevent injury from splinters. This is important when constructing a traditional roof, where timber is sawn and not planed.

Goggles

You would need to wear goggles for a range of carpentry tasks: for example, using a circular saw to cut roof joists to length.

Anything else?

Overalls provide some protection against the dust and keep your clothes clean. Simple safety glasses may be used for some types of machining.

BTEC **Assessment activity 10.3** P5 P6 M2

You are working on fitting a floor, but want to make sure that you work safely. You have two tasks to perform:

- laying chipboard flooring
- fixing new skirting boards.

1 Identify the PPE and safe working practices used to perform these carpentry tasks. **P5**

2 Explain why you selected the PPE and the safe working practices for the above two tasks. **P6**

3 Justify your choice of PPE and safe working practices to minimise health, safety and welfare risks for the two tasks. **M2**

Grading tips

1 To achieve **P5** you will first need to look at the type of work that is being undertaken and assess the hazards associated with this. Then you will need to decide what working practices and PPE is appropriate for reducing the risks while undertaking this work.

2 To achieve **P6** you need to describe the reasons why you picked these particular practices and PPE.

3 To achieve **M2** you need to set out the the advantages, disadvantages and benefits of each particular piece of PPE and safe working practice.

3. Safe working practices for carpentry tasks

3.1 Marking out

To carry out the assessment for this unit, and to achieve the grading criteria, you will need to know about marking out, producing and using setting out rods. Look at Unit 8 for information about how to do this.

3.2 Joints

You have already read about the different types of joint in Unit 8 (pages 76–77). Take some time to re-read this section now, making sure you cover all the following joints:

- housing
- through/corner halving
- tee halving
- through/corner bridle
- through/haunched mortise and tenon
- dovetail.

3.3 Carpentry task

Hanging doors: Side-hung using butt hinges

A typical door is hung on two or three hinges, depending on the weight of the door involved: for example, fire doors are a lot heavier than a domestic panelled door. Most doors are hung within a doorframe or lining set. This accepts the door and contains the doorstop or rebate that the door closes onto.

You need to mark out carefully using a pencil and tri-square, then cut timber from the the frame and door so that the hinges will sit flush with the surface of the timber. Hinges are commonly made from steel and secured with eight screws each.

Fixing simple mortise locks, cylinder night latches and lever handles

Doors and windows are normally fixed to frames using ironmongery. For security, a lock that inserts a bolt into the frame or keep when you turn a key is often used.

A mortise lock gets its name from the mortise that has to be cut into a door's rail to accept the lock. A cylinder night latch has a cylinder that holds the key-turning mechanism, and levels handles are those that attach to a lock, acting as a lever to open the latch that keeps a door shut.

Fixing picture rails, skirting boards, dado rails and architraves to masonry walls

Mouldings, a type of **second fixing**, are the finishing touches that are applied to a home. You must take great care when measuring, cutting and fixing these as they are visible finishes; joints need to be clean and tidy, accurate and well-finished.

A picture rail is used to hang pictures from on a wall; a dado rail breaks up high walls; a skirting board sits on the floor against the wall at skirting level. Architraves are used to cover the plaster joint between the doorframe and the finished plaster.

Look at how you mark out the hinge.

Key term

Second fixings – the secondary items, such as mouldings, that are mainly added after plastering has finished.

PLTS

Creative thinker ask your tutor questions to help you set our your frame

Reflective learner think about why any mistakes you make happened

Effective participator break your task down into manageable steps

Functional skills

Mathematics calculate the lengths accurately for your setting out rod

BTEC Assessment activity 10.4 P2 P4 P7 P8 P9 M1 M3 D1

You have been asked to produce a small-scale timber stud frame using a range of joints. You must clad this with plywood to the side and end face, and fix a small-scale skirting board and an architrave to include mitred joints.

1   Select the hand tools you would need to perform these carpentry tasks. **P2**

2   Select the materials required to perform this task. **P4**

3   Produce a setting out rod for cutting the mitres to the architrave and skirting board. **P7**

4   Set out and cut the two joints in the timber stud frame. **P8 P9**

5   Justify your use of hand tools and selected materials to minimise health, safety and welfare risks. **M1**

6   Produce finished work with all joints within 3 mm tolerance and square. **M3**

7   Produce finished work with all joints within 1 mm tolerance and square. **D1**

Grading tip

Look at the joints that you have practised and identify examples of where they can be used in finished timber products.

Samantha Hall

Estimator

Samantha works in the office of a construction company that produces timber framed houses for clients all over the UK. She works as an estimator providing prices against architects' drawings. She started this job after completing the First Diploma in Construction and then going on to a National Diploma in Construction at Level 3.

On the strength of the qualifications she had gained, she was taken on as an assistant detailer in the drawing office of the company. From this position she took on greater responsibilities and moved into the estimating department where she now works.

On a daily basis she meets architects, designers and contracts managers in the office and on site. She has found that with her experience as a detailer and knowledge of timber-framed construction she can quickly and efficiently sort out any problems that arise.

The company is keen to promote Samantha and she has been asked to run contracts. In this role she will have to recruit and train an assistant who will work alongside her. She is excited by this opportunity to train someone, as she feels this will be a way of repaying the company for the training she received as an assistant.

Samantha knows just where to find the right person for this position and contacts her tutor at the college she attended. She explains she is looking for someone who would be a suitable assistant and the tutor advises her that he will put forward three candidates for this position for Samantha to interview.

Think about it!

1. **How would you make sure you were one of these candidates?**
2. **What could you do now to improve your chances?**
3. **What documentation could you prepare to take to an interview?**

Just checking

1 When would you use a mitre box, and what would the angles in it be?
2 What does a nail punch do?
3 What two tools would you use to remove nails?
4 What is a skirting board?
5 What PPE would you wear to control noise?
6 What is the correct way to lift heavy objects?
7 Draw a sketch of a corner halving joint.
8 What materials are needed to hang and fit out a door?
9 What is the difference between a cylinder night latch and a mortise lock?
10 Where are lever handles fixed?

edexcel :::

Assignment tips

- Carpentry is all about thinking what you have to install into the space available. You will need to use measuring skills and be able to interpret drawn information.

- To get high marks, it will be essential that you can measure to length and work out angles accurately.

- Take care to check and re-check before you cut.

11 Exploring trowel operations

Bricks and blocks have been used for the construction of buildings for thousands of years. They are an attractive, economical and aesthetic building material, and have to be skilfully laid by a trained craftsperson – the bricklayer.

A bricklayer's skills are crucial in providing us with high quality homes to live in, suitable buildings to work in and many other structures that feature in our everyday lives. In this unit you will start to explore the manual skills needed to perform different trowel operations, understanding the hand tools that are required when laying bricks and blocks, how to safely achieve these tasks and the different factors that influence the quality of the completed job.

The technical skills of the bricklayer extend to the knowledge and understanding required to correctly set out brickwork in a range of 'bonds' or patterns. In this unit you will look at different bonds and how to achieve a quality outcome in the patterns you produce.

Learning outcomes

After completing this unit, you should:

1   know the hand tools and materials commonly used to perform brickwork and blockwork tasks

2   understand the important health, safety and welfare issues associated with brickwork and blockwork tasks

3   be able to apply safe working practices to set out and construct solid brick and block walling to given specifications.

Assessment and grading criteria

This table shows you what you must do in order to achieve a pass, merit or distinction grade, and where you can find activities in this book to help you.

To achieve a **pass** grade the evidence must show that you are able to:	To achieve a **merit** grade the evidence must show that, in addition to the pass criteria, you are able to:	To achieve a **distinction** grade the evidence must show that, in addition to the pass and merit criteria, you are able to:
P1 identify the hand tools used to perform brickwork and blockwork tasks **See Assessment activity 11.1, page 118**	**M1** justify the safe use of hand tools and materials to minimise health, safety and welfare risks **See Assessment activity 11.4, page 128**	
P2 select the hand tools used to perform brickwork and blockwork tasks **See Assessment activities 11.1, page 118 and 11.4 page 128**		
P3 identify the materials used to perform carpentry tasks **See Assessment activity 11.2, page 120**		
P4 identify the materials used to perform brickwork and blockwork tasks **See Assessment activity 11.4, page 128**		
P5 identify the PPE and safe working practices used to perform brickwork and blockwork tasks **See Assessment activity 11.3, page 125**	**M2** justify the use of appropriate PPE and safe working practices to minimise health, safety and welfare risks **See Assessment activity 11.3, page 125**	
P6 explain the selection of the PPE and safe working practices to be used in brickwork and blockwork tasks **See Assessment activity 11.3, page 125**		
P7 produce a gauge rod for setting out brick and block walls **See Assessment activity 11.4, page 128**	**M3** produce finished work with bricks in line to ±3 mm, plane face deviation ≤3 mm and corners plumb to within ±5 mm per metre height **See Assessment activity 11.4, page 128**	**D1** produce finished work with bricks in line to ±3 mm, plane face deviation ≤3 mm and corners plumb to within ±3 mm per metre height, plus clean face work and no interruption to bed joint continuity **See Assessment activity 11.4, page 128**
P8 identify common bonding arrangements used in the construction of solid brick and block walls **See Assessment activity 11.4, page 128**		
P9 produce solid brick and block walls to a given specification **See Assessment activity 11.4, page 128**		

How you will be assessed

This unit will be assessed by an internal assignment that will be designed and marked by the staff at your centre. Your assessment could be in the form of:

- presentations
- case studies
- practical tasks
- written assignments.

Stephen, 16 years old

A wall is just a wall, isn't it? This unit showed me otherwise, and made me look at the different sorts of brickwork and blockwork that are all around us. Different styles, different patterns, different purposes – and I did not realise that mathematics played such an important part in the setting out and construction of a wall, so it looks correct.

Laying bricks involves working outside in all weathers, which is just my sort of thing. So this unit has made me think hard about taking this trade on, as it has two aspects that I really enjoy – working with my hands and outdoor activity. I really enjoyed the way this unit took the theory we were taught in the classroom and put it into a practical context too, laying bricks in a pattern to produce walls.

Bricklaying is a solid trade, with an amazing history of craftsmanship behind it. I'd love to look into it more, maybe starting as an apprentice to begin with, so that I can learn from a qualified craftsperson.

Over to you!

- Would you want to work outside?
- Will you understand the application of mathematics in a practical way?
- How will you find out about further training?

1. Hand tools and materials commonly used in brickwork and blockwork tasks

Build up

Brick fix

Bricks have been made in the UK for hundreds of years, using clay extracted from the ground and furnaces to dry it out and harden the bricks.

These days, all bricks are manufactured in a standard size. Why do you think that is?

1.1 Hand tools

Walling trowel

This type of trowel is much larger than a normal pointing trowel (see the image below) as it has to lift a large volume of mortar to spread along a run of brickwork. Using a smaller trowel would take much longer and not be as efficient.

Pointing trowel

The pointing trowel is used for finishing off the joint formed between two bricks. This joint can be struck using a pointing trowel. A pointing trowel is also used where re-pointing is required on old brickwork joints.

Did you know?

The narrow end of the trowel blade is called the 'point'; the wide end, the 'heel'.

Pointing trowel

Jointing iron

A jointing iron is used to produce a half-rounded joint, which is slightly rounded to form a curved joint.

Jointing iron

Spirit level

A spirit level is used to ensure that you are laying your brickwork level and plumb. It is an essential tool for the bricklayer to ensure a quality job.

Spirit level

Builder's line and pins

A builder's line is used to ensure brickwork is straight and aligned, setting the level for the top of the brick when it is laid on the bed joint. Pins are placed in the corners and the line is stretched tight between them.

Builder's line and pins

Tingle plate

A tingle plate is a metal plate used to support the builder's line in the middle, to stop it sagging, which would mean that the brick laid would not be level or plumb.

Corner blocks

These do similar work to building line pins, but they are held at the corners of a wall as they grip the adjacent face of the brick, with the line running between the two corner blocks.

Corner blocks

Bolster chisel

There are two types of bolster chisel: a large one, used for cutting bricks, and a smaller one, used for detailed working, such as cutting recesses. Bolster chisels are manufactured from hardened tool steel, so they remain sharp for longer, and are available in standard widths of 100 mm, which fits the width of a normal brick or standard block. You should always have the hand guard in place, as this protects the fingers when you are gripping it in your hand and striking it with a hammer. Both types of chisel have hand guards on them to prevent you injuring your hands when you are using them. Never use a bolster without the hand guard in place. Any damaged hand guards should be replaced immediately.

Bolster chisel with hand guard

Club hammer

The club hammer or mash hammer is a heavier hammer than that used by a carpenter. It is used with a bolster chisel to cut bricks. It has a short handle and a heavy head for effective blows.

Club hammer

Brick hammer

One type of brick hammer is used to 'dress' brickwork when it has been cut to remove small pieces so a brick fits. This type consists of a short, hickory handle with a drop-forged head, which has a hammer at one end and a chisel at the other. Another type has a skutch comb for chasing or removing brickwork or blockwork.

Brick hammer

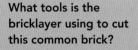

Assessment activity 11.1 **BTEC** P1 P2

You are working on a housing construction site, and your task is to set up and build the first **lift** off the foundations.

1 Identify the hand tools that you will need to take with you to perform this operation. **P1**

   Your tutor will give you a tool sheet that contains a series of tools from several trades.

2 Select the correct hand tools required to build a half-brick wall in stretcher bond, seven bricks long and seven bricks high. **P2**

Grading tips

1 For **P1** you will need to give a simple list of the correct tools that you will require to complete the task.

2 To achieve **P2** you must select the correct tools from the photo list. Look closely at each and decide carefully which are required for brickwork.

1.2 Materials

Common bricks

The common brick is an economical mass-produced brick, generally used where it will not be seen, such as a wall that is to be plastered or hidden. The common brick is produced in large quantities by the London Brick Company – hence its nickname, an LBC.

Facing bricks

A facing brick is the attractive finish brick used on the external skin of a house or commercial building. Facing bricks are produced in a variety of colours, surface finishes and textures. This final product that you see on a house or commercial building has to last a lifetime.

Engineering bricks

These are very heavy, dense and durable bricks, which are used in conditions that will be wet and damp, such as in a sewer, manhole or below the damp proof course in a home.

Solid blocks

These are solid, concrete blocks that are load-bearing and heavier than an insulation block (see below). They tend to have an open texture, ready for plastering and finishing.

Insulation blocks

These are lightweight blocks that contain a vast number of bubbles; these bubbles trap air, which is a great insulator. Insulation blocks often have a scratched pattern on the inside face, to help the plaster finish **key** to the blockwork.

Sand

Sand is mixed with water and cement to form mortar, which is used to bond the bricks or blocks together. The type of sand used will vary in texture and colour, depending on where it is quarried or extracted.

Cement

Cement is the ingredient in mortar that makes it set. It reacts with the added water to form a strong bond between the brickwork or blockwork and the mortar joint. Cement can be delivered to a construction site in two ways:

* bagged in 25 kg bags

* in a mortar silo.

Water is added via a pump at the base of the silo, mixing with the sand and cement to form ready-mixed mortar. Precautions must always be taken with cement as it can cause a chemical burn if in contact with moisture on the surface of the skin.

Lime

Lime is a product that was used before the invention of chemical **plasticisers** to make mortar more 'workable'. This means that it is more flexible and easier to move about as you spread it along a bed for the next row of bricks. In modern training workshops lime is added to the sand to act as a temporary setting agent. This allows the sand to set, allowing a joint to be finished and pointed. After the work has been assessed it can be taken down and recycled in a mill to produce fresh mortar.

Lime is an additive that can cause irritation, especially if absorbed into the moisture within the eyes. Lime is alkaline and, when mixed with water, can become corrosive, causing chemical burns to the surface of the skin and eyes. If it is inhaled, the moisture within your lungs will react with it, also causing an irritation. When in a training workshop, you should take great care to avoid lime mortar getting into your eyes and onto your face, and you should always wear eye protection.

> ## Key term
>
> **Key** – a scratched pattern in the surface of the block that makes the plaster coat adhere to the concrete block. To key means to adhere to.

> ## Key terms
>
> **Plasticisers** – modern chemicals that allow grains of sand to slide past each other easily
>
> **Workability** – the quality of being flexible and easy to move about

Water

Water is the ingredient that is used to mix the sand and cement, or the sand and lime, together. You must take water from a tap to ensure that it is of good quality and has no contaminants that could affect the strength of the mortar.

PLTS

Reflective learner ask your tutor for feedback on the list you make

Functional skills

Mathematics work out how many bricks you will need in total

BTEC Assessment activity 11.2 P3

You have been asked to carry out two trowel operations on a house building project:

- create an external skin to a house
- build an internal, load-bearing solid wall.

Identify the materials that would be used for each of these operations. P3

Grading tip

To achieve P3 you need to think about the two different areas where the materials will be used, and identify the correct materials from the appropriate section of the specification.

2. Health, safety and welfare issues for brickwork and blockwork tasks

2.1 Health, safety and welfare issues

Maintenance of a clean and tidy work space

A work environment is tidy when materials are stacked neatly ready to use and the area is clear of obstructions, particularly where people will be constructing walls.

A clean and tidy work area reduces the likelihood of an accident from a slip, trip or fall. For more information look at the section in this unit on slips, trips and falls.

Identification of hazards associated with given tasks

There are many hazards associated with brickwork and blockwork. The major ones are:

* working at height
* working with cement
* contact with moving machinery.

Identifying hazards comes with experience of working on construction sites. No two construction sites are the same as each has a unique design with its own hazards. You should always undertake a site induction when you first visit a site so that you are made aware of all of the hazards on site, such as someone working above you.

You should carry out a simple risk assessment before you start any given task. This will make you think carefully about the task you are about to undertake, and help you identify and control any hazards connected with this work.

Use of PPE to minimise risks from identified hazards

PPE or personal protective equipment is always the last resort when trying to control the risk from a hazard down to an acceptable level. PPE should always be assessed for its suitability, its fit and its performance in use. PPE must therefore:

* be worn correctly in order for it to work
* be maintained and looked after
* be stored correctly.

⚠ Cross reference

See Unit 2 pages 31–40 for more about risk assessment and controlling hazards.

2.2 Hazards

Slips, trips and falls

Slips, trips and falls can occur anywhere in a construction environment, especially where brickwork and blockwork are being carried out. If the brick and block waste from cutting is not cleared away, this can easily cause a slip or a trip. To prevent this, there should be a clearly defined cutting area available for all to use, or a block cutter placed in a waste management area. Everyone must be diligent about trip hazards and clean up their own work areas.

Cuts and injuries caused by tools and instruments

Bricklaying and pointing trowels have a sharp edge and a point, which can cause cuts to the hands if handled incorrectly. Cutting bricks is a task during which accidents can easily occur. To prevent accidents, always make sure that the bolster is sharp and used correctly, and that it has a hand guard fixed to it. If you slip with your hammer, the guard prevents you hitting the hand that you are holding the bolster with. Always make sure that the ends of bolsters you have 'rounded off' are ground back to their original condition.

Abrasive materials

Sand is a naturally abrasive material; we use it on sandpaper to wear timber products smooth. Sand is the main constituent of mortar, and can cause irritation to the eyes and mouth. Sand is usually wet or contains moisture, which prevents it being blown by the wind and causing harm and injury to the eyes. You should always wear eye protection when mixing sand to make mortar.

Lime

Lime can cause irritation to the lungs if it is inhaled. When mixed with water, it can cause chemical burns, especially to the skin and eyes. When working with lime, you should take great care to make sure that lime, or the mortar containing lime that you will use when training, does not get into your eyes or onto your face, and you should make sure you have appropriate protective clothing and eye protection.

Cement

The main hazard from using cement is a chemical burn. Cement is an alkaline material and reacts with water in a process known as hydration. During this process it gives off heat, which can cause damage to the skin and leave a cement burn mark. Cement dust that enters the eyes must be immediately washed out, as it will react with the moisture in the eye. You should take care when placing cement in the mixer so the wind does not get hold of it and blow it into your face and eyes.

Falling objects

When working at height, such as when constructing a first floor lift, bricklaying and building materials must be placed on scaffolding. In these situations, hazards include materials and people falling from the scaffold. The risk from falling materials is increased by:

- a lack of brick guards fitted to handrails to prevent materials falling through
- materials being stacked too high
- illegally adapting or removing pieces of the scaffolding.

There are strict safety procedures that you must follow when working on scaffolding in order to prevent people working on it from falling from height, which is the major cause of fatalities in the UK construction industry.

Untidy work area

An untidy work area can be an accident waiting to happen. A workplace that is covered in waste materials is obviously hard to walk over without tripping. Operatives must keep the work area clear of waste materials and quickly transfer these to a waste skip for disposal.

Musculoskeletal injuries resulting from lifting and moving heavy loads

Brickwork and blockwork operations often involve lifting heavy objects, such as concrete or steel lintels that bridge an opening, which have to support the loads placed on them. Before manually handling any object, you must:

- assess the weight of the building material you are about to move
- ascertain where its centre of gravity is
- decide whether a mechanical lifting means would be safer
- decide if it needs more than one person to lift it
- if safe, use kinetic lifting to lift the building material.

On construction sites, the risk of injury caused by lifting heavy objects can be minimised by using a mechanical telescopic handler. This piece of equipment has an extendable arm, which can be directed to place heavy materials right next to where they are required. This equipment saves the manual handling of heavy materials and the associated risk.

Did you know?

The HSE produce some startling statistics on the amount of days lost to the injury from back pain. Take a look at their website to find out more.

Key term

HSE – The Health and Safety Executive.

How do these boots protect the wearer?

2.3 PPE

Safety boots

These are essential, as the weight of a brick or block can easily break a toe or bone within your foot if dropped. Similarly, if you slip while carrying materials, they can fall and drop onto your feet. To protect you, a steel toe cap is built into the safety boot that resists the crushing or impact damage. A steel plate is also built into the base of the boot, to prevent penetration injuries to the sole of the foot.

Hard hat

As the name suggests, a hard hat is a helmet made of a hard, resistant plastic. There are several things about a hard hat that protect you.

- The hat has a peak to bounce any falling material away from your face.
- There is a sweatband to prevent the hat from slipping during movement.
- The helmet is adjustable for correct fitting.
- There are vents in the top to let air circulate around your head.

Wearing a hard hat on a construction site is mandatory, and suitable notices need to be posted around the site to inform all operatives of this requirement.

Activity: Hard hats

A hard hat is not a hat for life. Why is this? Find out when you should change a hard hat for a new one.

How does a hard hat protect the person wearing it?

High visibility jacket

A high visibility jacket or vest is essential so you can be seen on a building site where people, heavy machinery and vehicles come into close contact. High visibility strips are made of a reflective fabric that can be sewn onto clothing; at night, these strips reflect light from, for example, headlights, enabling you to be seen. T-shirts, coats, and trousers with these strips are available.

Activity: High visibility colours

There are two different coloured high visibility vests in use.

- **What are the colours used?**
- **Why is there more than one colour?**

Hand protection

The main type of hand protection for general labouring on site is a rigger's glove. A bricklayer will often wear an orange, non-slip glove that adds to the grip around a brick and prevents the skin coming into contact with the cement in the mortar. For someone working with blocks, gloves prevent abrasion injuries and can help to give grip when carrying heavy materials.

Goggles

You should always wear a pair of goggles when cutting bricks or blocks. Goggles are constructed so that they cover the whole of the eye and prevent any material from entering.

Look at how these goggles would wrap right around the eyes

Anything else?

You need to wear overalls to keep your clothes clean and to prevent mortar splashes from contaminating your clothes with cement.

BTEC · Assessment activity 11.3 · P5 P6 M2

You have been assigned to a team carrying out the following tasks:

- laying a row of facing bricks on a first floor scaffold
- cutting blockwork to size.

Your boss asks you to take special responsibility for health and safety, and to draw up a list of information for the team.

1 Identify the PPE and safe working practices used to perform the tasks. **P5**

2 Explain why you selected the PPE and the safe working practices for the tasks. **P6**

3 Give good reasons for your choice of appropriate PPE and safe working practices to minimise health, safety and welfare risks for the tasks. **M2**

Grading tips

1 To achieve **P5** you will need to correctly name the PPE required for the two tasks. The second part asks for the safe working practices to be identified and explained in performing the stated operations.

2 To achieve **P6** you need to explain why you chose that particular PPE and safe procedure by describing the process you went through.

3 To achieve **M2** you need to answer in more depth, adding to the answer for **P6** by outlining the reasons why you have chosen the PPE and stated methods.

PLTS

Independent enquirer explore the different safety issues for different team members

Functional skills

ICT use a computer to write up your findings

3. Safe working practices to set out and construct solid brick and block walling

Key terms

Bond – a technique of laying bricks on top of one another in a pattern, so that one brick always overlaps another.

Stretcher – a brick laid so its long side runs parallel to the face of the wall.

3.1 Half-brick walling

Half-brick walling is used in facings and is jointed on one side to straight lengths in **stretcher bond**. Stretcher bond is the pattern that can be seen where each brick is bonded half a brick on another to provide a strong wall. The diagram below shows a half-brick wall that is 102.5 mm wide, built in stretcher bond and jointed on the facework side using a jointing iron.

> Can you see how the bricks are bonded in their length?

3.2 One-brick walling

This type of walling is jointed on one or both sides and can be made in English bond and Flemish bond. You will need to know how one-brick walling is used to build straight lengths in facings.

One-brick walling is 215 mm wide and is made up of two 'leaves' of bricks.

English bond has one **course** of stretchers and one course of **headers**, which alternate on each course to form a pattern and create a very strong wall.

Flemish bond consists of alternate stretcher and headers for each course, forming a unique pattern on the face.

> Can you see both leaves of bricks making up this one brick wall?

Key terms

Course – a row of bricks or blocks.

Header – a brick laid so that its short side is parallel to the face of the wall.

> Can you identify a course of headers and a course of stretchers in this English bond wall?

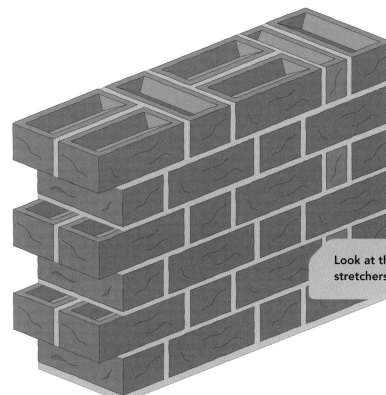

Look at the pattern formed by the alternating stretchers and headers in this Flemish bond wall

3.3 Block walling

Block walling is constructed in the same way as brick stretcher bond, but the blocks used are obviously much larger. You might need to construct a straight length of block walling in stretcher bond.

3.4 Brickwork

Gauge rods

A gauge rod can be used to gauge the correct height of the courses in a wall. A gauge rod for brickwork is a simple piece of timber with saw cuts placed in it 75 mm apart for a string line to fix in. The cuts are 75 mm apart to mark out the height of a brick (65 mm) plus a joint (10 mm). When lines are stretched horizontally between two gauge rods, you can set out the courses in a brick wall by following the lines.

3.5 Blockwork

Gauge rods

For blockwork, gauge rods are prepared with cuts 225 mm apart. The same gauge rod used for brickwork can also be used for blockwork; every third saw cut will show where a block course should be laid, because the block course is equal to three courses of brickwork (75 mm × 3 = 225 mm).

BTEC Assessment activity 11.4 · P2 P4 P7 P8 P9 M1 M3 D1

You have been asked to produce a small sample of brickwork or blockwork as indicated on the drawing provided by your tutor. The panel is to be jointed using a jointing iron.

1 Select the hand tools you would need to perform these tasks. **P2**

2 Select the materials required to perform these two tasks. **P4**

3 Produce a gauge rod for setting out brickwork and blockwork walls. **P7**

4 Identify three common bonding arrangements used in solid brick and block walling. **P8**

5 Produce the two samples of walls in accordance with the drawing provided. **P9**

6 Justify your use of hand tools and selected materials to minimise health, safety and welfare risks. **M1**

7 Produce finished work with bricks in line to ±3mm, plane face deviation ±3mm and corners plumb to within ±5mm per metre height. **M3**

8 Produce finished work with bricks in line to ±3mm, plane face deviation ≥3mm and corners plumb to within ±3mm per metre height, plus clean face work and no interruption to bed joint continuity. **D1**

Grading tips

1 To achieve **P2** you will need to demonstrate that you can select the hand tools for each joint correctly by verbal questioning and the completion of an observation record.

2 To achieve **P4** you must correctly select the right materials for completing each joint. Again, the above evidence method can be used.

3 To achieve **P7** you will need to draw out a gauge rod correctly for the dimensions given to you in the brief.

4 To achieve **P8** you will need to correctly identify the bonding arrangements in the types of walling drawings you have been given.

5 To achieve **P9** you must complete the two samples of walls.

6 To achieve **M1** you need to explain your reasons for choosing the hand tools and materials for this task.

7 To achieve **M3**, which is a quality criterion, you need to produce work that is accurate within the stated tolerances.

8 To achieve **D1** your work needs to be completed to a very accurate level, taking great care with detail, level and plumb.

Bricklayer's labourer

Mohammad has been working as bricklayer's labourer within a '2+1' gang for the past three years. This is hard manual work involving stacking out materials for the two bricklayers, mixing the mortar they require on site to lay bricks and dealing with insulation materials and clips when required.

He realises that he doesn't want to keep doing this for the rest of his working career and needs to act fast. He needs to change and train for a career in brickwork and eventually site management where the rewards are greater and the work is far more technical, which will challenge his abilities.

Mohammad wants to start on the NVQ Trowel Operations Level 2 at a local technical college one day per week to learn the theory and application in a practical workshop. The rest of the time he would be working but gathering evidence of other work that he has done on site and building up a portfolio.

The grade C GCSE he gained in English at school will help him with the written work in his studies. The other two bricklayers say they will help him with starting to work on the trowel to improve his skills, and as he gets better they will increase his portion of the wages as he becomes productive in laying bricks and blocks.

Mohammad now considers that he has a great opportunity to eventually become a site manager who is responsible for the whole building project from start to finish. This is a clear goal he wants to achieve.

Think about it!

1 What can Mohammad do to achieve his ambition?
2 What further qualifications will he need to become a site manager?
3 How long could this take him?

Just checking

1 What is a header?
2 What is the length of a brick?
3 What is a brick laid parallel to the face called?
4 What does bricklaying mortar contain for training purposes?
5 Is cement harmful? If so, why?
6 Draw a panel of English bond.
7 Draw a panel of Flemish bond.
8 What should be attached to a brick bolster to prevent hand injuries?
9 What does a tingle plate do?
10 How do a pointing trowel and a walling trowel differ?

edexcel :::

Assignment tips

- Laying bricks and blocks is all about manual dexterity, so practice makes perfect.
- Always use a level to continually check your work for level and plumb.
- Stand back from your work and have a look at it with your eye to check for quality.
- Keep your work area tidy and clean using a mortarboard.

12 Performing blockwork operations

Blockwork plays an important role in the modern construction industry. Blocks are often used in preference to bricks when building and forming walls because they are larger than bricks so fewer are required in any given structure, saving labour, time and money.

Blocks manufactured from concrete are a 20th century invention and have replaced the inside skins of hollow walls with a material that is load-bearing and quick to lay. With the advent of a chemical process that aerates the concrete, air can be trapped within a block to make it a good insulator against heat loss.

Block technology has advanced into the thin joint masonry system whereby larger blocks are glued together using a cement-based adhesive to form a quick, strong, high-insulation wall.

Blockwork tends to be covered up with plaster finishes and so does not have to be finished with the quality pointed face that facing brickwork requires as it is very rarely seen, except in areas such as a garage for a house.

Learning outcomes

After completing this unit you should:

1 know the hand tools and materials commonly used to perform blockwork tasks

2 understand the important health, safety and welfare issues associated with blockwork tasks

3 be able to apply safe working practices to the setting out and construction of corners and junctions in solid block walling to given specifications.

Assessment and grading criteria

This table shows you what you must do in order to achieve a pass, merit or distinction grade, and where you can find activities in this book to help you.

To achieve a **pass** grade the evidence must show that you are able to:	To achieve a **merit** grade the evidence must show that, in addition to the pass criteria, you are able to:	To achieve a **distinction** grade the evidence must show that, in addition to the pass and merit criteria, you are able to:
P1 identify the hand tools used to perform blockwork tasks **See Assessment activity 12.1, page 136**	**M1** justify the safe use of hand tools and materials to minimise health, safety and welfare risks **See Assessment activity 12.4, page 140**	
P2 select the hand tools required to perform given blockwork tasks **See Assessment activity 12.4, page 142**		
P3 identify the materials used to perform blockwork tasks **See Assessment activities 12.1, page 136 and 12.2, page 137**		
P4 select the materials required to perform given blockwork tasks **See Assessment activity 12.4, page 142**		
P5 identify the PPE and safe working practices used to perform blockwork tasks **See Assessment activity 12.3, page 140**	**M2** justify the use of appropriate PPE and safe working practices to minimise health, safety and welfare risks **See Assessment activity 12.3, page 140**	
P6 explain the selection of the PPE and safe working practices to be used in given blockwork tasks **See Assessment activity 12.3, page 140**		
P7 identify the correct bonding arrangement to be used in the construction of solid block walling **See Assessment activity 12.4, page 142**	**M3** produce finished work with blocks in line ±3 mm and face plane deviation ≤3 mm **See Assessment activity 12.4, page 142**	**D1** produce finished work with blocks in line ±3 mm and face plane deviation ≤3 mm and with corners plumb to within ±3 mm per metre height **See Assessment activity 12.4, page 142**
P8 set out block walling details to given dimensions with guidance and supervision **See Assessment activity 12.4, page 142**		
P9 produce corners and junctions in solid block walling to given specifications **See Assessment activity 12.4, page 142**		

How you will be assessed

This unit will be assessed by an internal assignment that will be designed and marked by the staff at your centre. Your assessment could be in the form of:

- presentations
- case studies
- practical tasks
- written assignments.

Jan, 16 years old

Before this unit, I hadn't realised that many houses are constructed using blockwork, as you cannot see where blocks have been used within the construction. When I did this unit, I found the blocks so much bigger than the bricks I'd been using in Unit 11. They were heavier and much harder to handle, and at first I thought I'd never do it!

As the unit went on though, I really got the hang of it. Because of the way the blocks are, you have to get the first course level; if you don't, it will affect the rest of the wall that you are constructing. I learned that getting it right first time will save me effort in the long run.

It was good to produce blockwork models, which are going to be plastered. I had to really work on my accuracy, making sure my blockwork was plumb and level, both for the plasterer to get it smooth and for the joiner to put the first floor joists in.

The blockwork doesn't require pointing like the brickwork and so I don't have to worry about the finish produced in the face, which is great!

Over to you!

- Will you enjoy working outside in this industry?
- What standard will be required on a site?
- What other skills will you need to develop?

1. Hand tools and materials commonly used in blockwork tasks

Build up

Blocked in

Blocks are a faster method of producing walls than bricks, as you can fit a large number of bricks within the space that a block occupies. How many bricks fit within one block?

⚠ Cross reference

Read Unit 11, pages 114–117 for descriptions, details and photos of these tools.

Bolster chisel

1.1 Hand tools

When constructing blockwork, you will need to know what these tools look like and how they are used:

- walling trowel
- pointing trowel
- jointing iron
- spirit level
- builder's line and pins
- tingle plate
- corner blocks
- club hammer
- bolster chisel
- brick hammer.

Walling trowel

A walling trowel is essential for laying blocks as you need a large amount of mortar on the trowel to bond the ends, because they are taller. Because of their size, you can lay these more quickly than brickwork.

Pointing trowel

Occasionally blocks may require pointing with the pointing trowel, but this is very rare. Areas that may require pointing would be in places like sports facilities, where a weather struck or flush struck joint may be required if the walls are to be painted over as a finish.

Jointing iron

Fair-faced blockwork can be finished in 'half-rounded' using a jointing iron: for example, on a garage where it is more economical to use blockwork inside than facing brickwork. You would use the same tool and process as used for jointing brickwork.

Spirit level

With blocks you need to use a longer spirit level, which makes the job of vertical plumbing easier. A spirit level should have two bubbles: one for horizontal and one for vertical.

Spirit level

Builder's line and pins

The builder's line for blockwork will be set at 225 mm intervals, as this is equivalent to one block and a bed joint. The line and pins are used in exactly the same way as in brickwork.

Club hammer

A club hammer is heavy enough to cut blocks. When you use a club hammer with the bolster chisel to cut blocks, you will need to make several small cuts along the cut line on the block to achieve a clean cut.

Brick hammer

The comb on a brick hammer can be used to trim and shape a cut block so that it sits correctly within the gap.

Gauge rods

A gauge rod is simply a piece of timber or plywood with saw cuts placed every 75 mm. You set the rod against a wall to provide the gauge from the **datum** vertically up the wall.

For blockwork, every third saw cut will show where a block course should be laid, because the block course is equal to three courses of brickwork (75 mm × 3 = 225 mm).

For blockwork, you also need to know about building profiles.

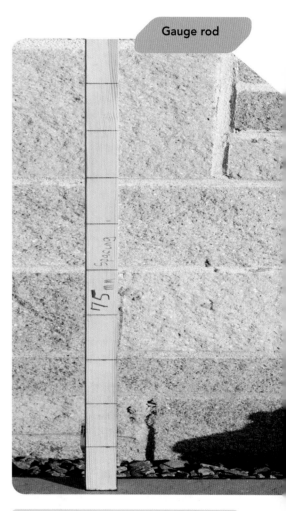
Gauge rod

Building profiles

A building profile fits on the corner part of brickwork or blockwork and is fixed using cramps, which fit into the vertical joints. This corner post then acts as a template to build to, as it is both vertical and plumb. A string line can be attached to it, making the job of laying blocks that much faster.

Key term

Datum – a level given which all other levels relate to.

BTEC **Assessment activity 12.1** P1 P3

You are working on building a solid block internal wall on the ground floor of a house.

1. Identify the hand tools that you will need when performing this operation. P1

The site foreman has asked you what materials you require for this task on site.

2. Write out a simple list for the foreman. P3

Grading tips

1 To achieve P1 you will need to list the correct tools that you will to complete the task.

2 To achieve P3 you need to list the materials that you will require to build this internal block wall.

1.2 Materials

Lightweight aerated blocks

These light insulation blocks are very thermally efficient, as they have an air-entraining agent introduced during manufacture. This agent makes the concrete foam up and capture air bubbles, which set in the concrete. Trapped air is an excellent insulator, and so is the block.

Dense solid blocks

Dense solid blocks are available in 3.5 N (Newtons) and 7 N strengths. They are used in areas of load bearing, or where sound proofing is an important factor. By their nature they are heavier than lightweight blocks, so you must take care with manual handling.

Hollow blocks

Hollow blocks are larger-width blocks that provide the same strength but at a more economical price and with less weight. The hollows can be filled with concrete and steel rods for added strength.

Fair-faced and keyed blocks

A fair-faced block has one face that is clean, tightly closed and suitable to be painted or for a self-finish. It is often classed as 'paint grade'. External fair-faced blocks can have a secondary finish placed on them, such as artificial stone.

Keyed blocks have a distinctive design, with a zig-zag face. The surface has been scratched to allow the plaster or cement render to bond better to the block's surface.

Why do you think aerated blocks provide good insulation?

Activity: Identifying blocks

The following block manufacturers produce many of the blocks mentioned earlier. Investigate one of their websites and identify the different types of blocks available.

- Tarmac
- Celcon
- Hanson

Lime

It is very rare that you would add lime when mixing mortar for blocks. Apart from training purposes, the only instance where lime is added would be in the restoration of a listed building, where new materials have to match the existing ones.

Sand, cement and water

Other materials you will need to know about for this units are sand, cement and water.

Cross reference

Look at Unit 11, pages 119–120, for more information on sand, cement, lime and water.

Assessment activity 12.2   P3

BTEC

You have been asked to carry out the following blockwork operations:

- creating an internal skin to a house
- building a party wall between houses.

Identify the materials you would need for each task. **P3**

Grading tip

For **P3** you need to think about the two different areas where the materials will be used for each and identify these. Your research from the block activity will be a great help here.

PLTS

Self-manager show initiative in finding the information you need

Functional skills

ICT use DTP to present your findings in an interesting way

2. Health, safety and welfare issues for blockwork tasks

2.1 Health, safety and welfare issues

Health, safety and welfare issues for the construction industry as a whole are covered in Unit 2 of this book. Look back at this unit to remind yourself of these. You should also refer to Unit 11 (pages 121–125), which covers health and safety issues for brickwork and blockwork, including the correct types and uses of PPE.

Maintenance of a clean and tidy work space

Blocks are larger than bricks, so present a much bigger trip hazard if you leave them untidy. Blocks should be stacked no more than 900 mm high, immediately next to where they are required. You should sweep up any waste materials from cutting and clear them away, to be disposed of safely off site or used as hardcore.

Identification of hazards associated with given tasks

Block laying has a number of hazards associated with it.

- Larger-width blocks, when wet, can weigh over 30 kg, which presents a hazard when lifting into position.

- Working at height on scaffolding can lead to falls.

- Walls that are built too high can be overturned by the wind.

You should undertake a simple risk assessment for any construction activity to identify any hazards associated with the work.

Use of PPE to minimise risks from identified hazards

Hand protection is essential when you are lifting dense concrete blocks. They tend to have a rough, abrasive finish to their surface, which can cause cuts to the hands and abrasion burns when manually handling them. Always use mechanical means and move them in bulk, in the packs they have been delivered in.

On a construction site you should also always wear a hard hat, safety boots and a high-visibility jacket. For more information about PPE, look at Unit 11, pages 124–125.

> ⚠ **Cross reference**
>
> See Unit 2 pages 31–40 for more about risk assessment and controlling hazards.

2.2 Hazards

Slips, trips and falls

When you are working with heavier blocks, you may be more likely to slip or trip. You should take great care with preparing the work area so that a clear route is left for materials and people to pass through and work in. Cut blocks should be removed from scaffolding using a rubbish chute that discharges into a covered skip. Many blockwork operations are conducted at height where there is the risk of falling. Safety measures need to be in place to prevent falls occurring.

Cuts and injuries caused by tools and instruments

If a block cutter is not used correctly, it can cause crush injuries. A masonry saw is often used to cut blocks on site. This has a petrol-driven rotating blade that cuts through the blocks using abrasive materials, so when using a masonry saw, you must use dust and eye protection.

How could a block cutter cause a crush injury?

Cutting blocks

Blocks can be cut using a petrol driven saw, a block splitter or a bolster chisel and hammer. Whichever method you choose, you must take the necessary precautions by using the correct procedures and PPE.

Abrasive materials

Sand, which is the main constituent of mortar, can cause irritation to the eyes, skin and mouth if it comes into contact, so you must wear eye protection when mixing mortar. The rough texture of the concrete within blocks can cause abrasion to the hands if you handle them without gloves.

Cement

Concrete in whole blocks is relatively harmless but, when a block is cut using a petrol saw, the cement is broken up into a fine dust, which can enter the lungs and eyes, causing irritation. You must wear suitable protective goggles and dust masks when conducting this operation.

 Cross reference

Look at Unit 11, page 123 to recap the risks of falling objects and having an untidy work area.

Musculoskeletal injuries resulting from lifting and moving heavy loads

Laying blocks that have been left out in the rain often involves lifting a considerable weight, especially if they are 200 mm-thick party-wall blocks. These can weigh in excess of what is safe to lift individually, so you may need two people to handle each block into position. With all blocks you should:

* assess the weight of the block before lifting

- ascertain where its centre of gravity is

- decide whether a mechanical lifting means would be safer

- decide if it needs more than one person to lift it

- use kinetic lifting to lift the block into position.

2.3 PPE

When performing blockwork operations you should always wear:

- safety boots
- hard hat
- high-visibility jacket
- gloves
- overalls.

Cross reference

Look at Unit 11, pages 124–125. for more information about these items of PPE.

What items of PPE is this person wearing while building a block wall?

PLTS

Independent enquirer make a list of criteria your PPE needs to meet

Functional skills

English use persuasive language to justify your choices

BTEC Assessment activity 12.3 P5 P6 M2

You have been assigned to a project laying some foundations for a commercial building. This will involve:

- laying dense foundation concrete blocks below ground level

- cutting blocks using a masonry saw.

Before you start, you need to look at the health and safety issues involved.

1 Identify the PPE and safe working practices you will need to perform the tasks. **P5**

2 Explain why you selected the PPE and the safe working practices for the above two tasks. **P6**

3 Justify your choice of appropriate PPE and safe working practices to minimise health, safety and welfare risks for the two tasks. **M2**

Grading tips

1 To achieve **P5** you will need to correctly identify the PPE required for the two tasks, then describe the safe working practices you will need to follow for each.

2 To achieve **P6** you must outline why you chose each item of PPE and each safe procedure.

3 To achieve **M2** you will need to add depth to your answers, giving the reasons for your choices.

3. Safe working practices to set out and construct corners and junctions in solid block walling

3.1 Setting out

Calculating lengths of walling

Correctly calculating lengths of walling involves taking a dimension from a drawing and calculating how many blocks will be required to build it. You will need a small amount of mathematics to set out the blockwork!

To find out how many blocks will be needed to build one course of a wall, you need to divide the length of the wall by the length of the block plus the width of joint between the blocks. Blocks are 440 mm long, and joints should be 10 mm wide; this means that the length of a block and a joint is 450 mm. To find out how many 450 mm blocks and joints will be needed to build a wall, divide the wall's total length in mm by 450.

Look at this blockwork in stretcher bond

Bonding and lengths of walling

To allow the bonding of the blockwork (that is, to allow the next course to overlap the previous one by half a block), you may have to place a cut block somewhere within the completed wall. If the job does not work with full block sizes, you will need to consider this carefully to ensure that the wall remains fully bonded. In blockwork, bonding will always be in stretcher bond, as this is the only type of bonding that can be used with blocks to provide a strong wall.

Activity: Calculate

The architect has given you the following dimensions for a room: 5.40 m × 4.05 m. Calculate how many blocks this is per side.

Indents and toothings to walling

Indents are where a gap is left within a wall at every other course for an opposing wall to slot in at a later date. The gap left is the width of the wall that will join it. Toothings are blocks that stick out from a wall at every other course, ready to receive a wall at a later date.

3.2 Block walling in stretcher bond

Corners

When constructing corners in blockwork, you first build one each side of the building, then put up a string line or corner profiles to help you infill the courses between the corners, in order to complete the wall. This is done in lifts, which are normally 900 mm high.

Can you see how the blocks meet at the corner?

T-junctions

A T-junction is where an internal block wall meets with another wall and forms a 'T' shape. This joint cannot be straight vertically, as cracks would form at the corner of the wall when it is plastered. For this reason, the blocks must tie in and overlap with the external wall courses to form a strong bond.

Staggered junctions

Staggered junctions are the same as T-junctions, except that the wall is staggered with a free end, which is not built into the opposing wall. The strength of the free end is supported by its bonding to the floor and fixing at ceiling level.

Assessment activity 12.4

P2 P4 P7 P8 P9 M1 M3 D1

You have been asked to produce a small sample of blockwork as indicated on the drawing provided by your tutor. This sample will be used in the selection process for a modern apprenticeship.

1 Select the hand tools you would need to perform the given blockwork tasks. **P2**

2 Select the materials required to perform these two tasks. **P4**

3 Identify the bonding arrangement you are going to use. **P7**

4 Set out the sample panels in accordance with the drawings issued. **P8**

5 Produce a corner and a wall junction in 100 mm blockwork. **P9**

6 Justify your use of hand tools and selected materials to minimise health, safety and welfare risks. **M1**

7 Produce finished work with blocks in line ±3 mm and face plane deviation ±3 mm. **M3**

8 Produce finished work with blocks in line to ±3 mm, and face plane deviation ≤3 mm and corners plumb to within ±3 mm per metre height. **D1**

Grading tips

1 To achieve **P2** you will need to show that you can select the hand tools for the completion of the given blockwork tasks. This can be recorded on a witness or observation record as evidence.

2 To achieve **P4** requires that you correctly select the right materials for the blockwork tasks. It would be useful for a requisition order to be supplied here.

3 To achieve **P7** you will need to identify the correct bonding arrangement to use to produce the blockwork tasks.

4 To achieve **P8** you will need to correctly set out the model that you have been asked to build. You could self-assess the setting out,

then ask your tutor to check it and record it formally.

5 To achieve **P9** you must complete a blockwork corner with a straight wall and a junction within it to the given specification.

6 To achieve **M1** you must explain the reasons why you have chosen the hand tools and materials for the tasks.

7 To achieve **M3** which is a quality criterion, you will need a level of accuracy to produce the work within the set tolerances.

8 To achieve **D1** you will need to work very accurately, with great care for detail, level and plumb.

Paul Repper
Apprentice builder

Paul is a second-year brickwork student employed as an apprentice by a local construction company. He is currently only laying blockwork to internal walls on site, as this type of work does not have to be as accurate as facing brickwork. This is because it is mainly covered up with secondary finishes. The blockwork Paul builds must be vertically plumb and straight so that walls appear straight.

However, Paul would like the opportunity to practise and improve his facing brickwork so that he can complete his NVQ Level 2, as he is missing this part of the evidence.

Paul attends college one day a week. In order to gain the skills he needs he asked his tutor if he could build corners in face brickwork with a cavity and an internal skin of blockwork, until he achieved an acceptable standard of quality. To gain extra experience Paul is even working outside class time to improve the work on his facework model.

Paul told the site manager at work about the skills he practised at college and the manager gave him the opportunity to build a corner in facing brickwork, for one of the first lifts of the new homes the company is building. The site manager was very impressed with Paul's work. Paul is pleased with this as his work is on the show home that potential buyers view when they are thinking about buying a house on site.

Paul enjoys the blockwork he is doing during his apprenticeship, and is inspired to keep working hard when the effort he puts in at college brings new opportunities at work. Paul would like to become an assistant site manager for his company.

When he has an appraisal, Paul asks the training manager about BTEC Level 3 management courses. The training manager suggests that he needs at least five years of experience in the construction industry and suggests Paul attends college one day per week on the National Certificate in Construction and takes on a few management duties on site.

Think about it!

1 Can Paul use his site work as evidence for his NVQ Level 2?

2 Where can Paul find out about qualifications?

3 What steps can he take to develop both his practical and management skills?

Just checking

1 What is the difference between an aerated block and a solid block?
2 Do you joint blocks in the same way as brickwork?
3 What methods are available to cut concrete blocks?
4 What does a gauge rod do?
5 How do building profiles work?
6 What is a fair-faced block?
7 What are the major hazards in laying blocks?
8 What is the main type of bonding used in blockwork?
9 Why is a spirit level important in blockwork?
10 Would you use a pointing trowel in laying blockwork?

edexcel

Assignment tips

- Laying blocks requires that you continually check line and level, especially plumb as you don't want the wall to deviate and lean.

- Make sure that your work area is clean and tidy to avoid any trip hazards.

- Stand back from your work and have a look at it to check for quality.

- Use a mortarboard for your sample panels.

13 Performing brickwork operations

Bricklaying is a traditional craft that has been practised for thousands of years. It involves the ability and skill using manual dexterity to correctly lay a brick on a bed of mortar ensuring that it is level and vertical. Once finished a brick wall can look superb and add value to a high quality home.

Training to become a bricklayer may involve starting an apprenticeship with a local construction company where you will learn from on-the-job training and attending college at least one day per week. This is a great way to learn the trade and the skills required to produce quality work.

Learning outcomes

After completing this unit, you should:

1 know the hand tools and materials commonly used to perform brickwork tasks

2 understand the important health, safety and welfare issues associated with brickwork tasks

3 be able to apply safe working practices to the setting out and construction of brickwork to given specifications.

Assessment and grading criteria

This table shows you what you must do in order to achieve a pass, merit or distinction grade, and where you can find activities in this book to help you.

To achieve a **pass** grade the evidence must show that you are able to:	To achieve a **merit** grade the evidence must show that, in addition to the pass criteria, you are able to:	To achieve a **distinction** grade the evidence must show that, in addition to the pass and merit criteria, you are able to:
P1 identify the hand tools used to perform brickwork tasks **See Assessment activity 13.1, page 150**	**M1** justify the safe use of hand tools and materials to minimise health, safety and welfare risks **See Assessment activity 13.4, page 158**	
P2 select the hand tools required to perform given brickwork tasks **See Assessment activity 13.4, page 158**		
P3 identify the materials used to perform brickwork tasks **See Assessment activity 13.1, page 150 and 13.2, page 151**		
P4 select the materials required to perform given brickwork tasks **See Assessment activity 13.4, page 158**		
P5 identify the PPE and safe working practices used to perform brickwork tasks **See Assessment activity 13.3, page 154**	**M2** justify the safe use of appropriate PPE and working practices to minimise health, safety and welfare risks **See Assessment activity 13.3, page 154**	
P6 explain the selection of the PPE and safe working practices to be used in given brickwork tasks **See Assessment activity 13.3, page 154**		
P7 identify the correct bonding arrangements to be used in the construction of brickwork **See Assessment activity 13.4, page 158**	**M3** produce finished work with bricks in line to ±3 mm, face plane deviation ≤3 mm and corners plumb to within ±3 mm per metre height **See Assessment activity 13.4, page 158**	**D1** produce finished work with bricks in line to ±3 mm, face plane deviation ≤3 mm and corners plumb to within ±3 mm per metre height **See Assessment activity 13.4, page 158**
P8 set out brickwork to given dimensions with some guidance and supervision **See Assessment activity 13.4, page 158**		
P9 produce brickwork to given specifications **See Assessment activity 13.4, page 158**		

How you will be assessed

This unit will be assessed by an internal assignment that will be designed and marked by the staff at your centre. Your assessment could be in the form of:

- presentations
- case studies
- practical tasks
- written assignments.

Linda, 14 years old

I enjoy working with my hands and love a challenge, so brickwork has been great for me. At first it wasn't easy but, after persevering, I'm doing really well. I like the model work best. We get to use different bricks and pointing methods to produce different finishes and it feels good when you make something that looks neat and solid.

I was really pleased when my teacher approached me to say I was one of the learners he was putting forward for an apprenticeship place with a local company that's are keen to promote equal opportunities in construction. This has made me really motivated to do well and my model has been assessed as a distinction-level piece.

My mates used to give me a hard time for choosing brickwork but now they respect me when they see the quality of my work. And now I have the chance to train though the apprenticeship, which will give me a great opportunity to prove myself even more.

Over to you

- How will you obtain a training position with a company?
- What qualities should a bricklayer have?
- What skills will you need to improve?

1. Hand tools and materials commonly used in brickwork tasks

Build up

Brickwork by numbers

Bricklaying involves a certain amount of mathematics in calculating the correct number of courses and bricks per course, so quantities can be ordered for delivery to site and architect's dimensions kept to accuracy.

Is your level of maths sufficient for this?

What will you do to improve this?

1.1 Hand tools

When performing brickwork operations, you will need to know what these tools look like and how they are used:

- walling trowel
- pointing trowel
- jointing iron
- spirit level
- builder's line and pins
- tingle plate
- corner blocks
- club hammer
- bolster chisel
- brick hammer

Before reading on, look back at the descriptions, details and photos of these tools in Unit 11, pages 114–117, then read the additional information in this unit about how these tools are used in brickwork.

Walling trowel

The walling trowel is the tool that a bricklayer will use most often. The photograph here shows a bricklayer using a walling trowel to lay a course of bricks. Look at the bed of mortar that the builder in the photo has laid down with the walling trowel. It is worth spending more on this tool, as you will want to get the most out of the trowel in the long run.

Pointing trowel

There are many different types of joint used to point facing brickwork for which a pointing trowel can be used. The weather struck and reverse struck joints are different ways of using the light on a wall to make the joint appear different. This is achieved by angling the trowel forwards or backwards on the mortar joint.

A bricklayer using a walling trowel

Jointing iron

The jointing iron has two different-sized ends: most commonly one end is 15 mm and the other 12 mm. The jointing iron is run along the wet mortar, compressing it into the joint to form a concave pointing that is weatherproof.

Spirit level

This may well be the most expensive tool a bricklayer will buy, as it has to be accurate in order to keep the brickwork level and plumb. Always check the accuracy of your level to ensure that it has not become damaged, as this will affect the quality of the brickwork you produce. Normal bricklayer's levels start at 800 mm long. You can keep a small torpedo level in your tool bag for those awkward, small areas where the ordinary level would be too long to use accurately.

Builder's line and pins

A builder's line and pins are used to ensure the top course of bricks is straight and aligned. When building brickwork, you complete one corner first, then the second corner, and then use a builder's line between the two to work from. This line ensures that the brickwork between the two corners is straight and aligned properly. The pins are pushed into the mortar joints and the line stretches across the area you want to run a course along.

Corner blocks

Corner blocks hold a builder's line in place. They are positioned at the height of the course the bricklayer is working on, and mark where they should lay the bricks. The tension on the line holds the blocks in place until the course is finished and then the line is moved up to mark the next course.

Club hammer

The club hammer traditionally has a timber shaft made of hickory, which is a harder timber that resists splitting. However, the modern club hammer may have a graphite or synthetic shaft instead. The shaft is shorter than a carpenter's hammer to make it easier to control and more accurate when cutting bricks. The weight of the head is available in a series of light to heavy grades.

Gauge rods

A gauge rod has saw cuts placed at every 75 mm. For brickwork, this is the standard course for metric bricks where 65 mm plus a 10 mm mortar bed (65 mm + 10 mm = 75 mm). You set the rod against a wall to provide the gauge from the datum vertically up the wall.

Did you know?

The longer the level, the greater the accuracy it will give.

Building profiles

A building profile fits on the corner part of the brickwork or blockwork and is fixed using cramps which fit into the vertical joints. This corner post then acts as a template to build to as it is both vertical and plumb and a string line can be attached to it. This makes the job of laying the bricks or blocks that much faster.

Key terms

LBC – London Brick Company.

Impermeable – will not let water absorb into it or pass through it.

Assessment activity 13.1 **BTEC** P1 P3

You are working on building a solid one-brick garden wall 3 m long × 1 m high.

1 Identify the hand tools that you will need when performing this brickwork operation. **P1**

2 The materials are going to be supplied by the client. Provide a list for them to buy. **P3**

Grading tips

1 To achieve **P1** you will need to make a simple list of the correct tools that you will require to complete the task.

2 To achieve **P3** you need to create a simple list of the materials that you will require to build this small garden wall.

1.2 Materials

Common bricks

The **LBC** common is an economical brick that is used where it will not be seen, so it will not be used for facework. A common brick has a frog, which is the void in the base. Common bricks can be laid frog up, or frog down for economy. When laid frog up, they are much stronger as mortar fills the frog and forms a solid bond.

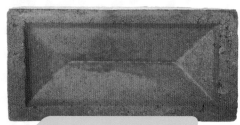

Frog up or frog down?

Engineering bricks

These are solid, very dense bricks. Because of the density, there are few pores of air left in the brick and it is almost impermeable to water. Engineering bricks are therefore used below the damp proof course where they are ideally suited.

What makes engineering bricks suitable for use below the level of a damp proof course?

Special bricks

A special brick is one that has a shape that is not rectangular like a normal brick. These are used to provide architectural details on a brick facade of a building. Sometimes a particular type of brick has to be specially manufactured for a specific brick project.

Did you know?

You can manufacture special bricks by cutting and epoxy-gluing bricks together.

Activity: Special bricks

Special bricks have many uses in detailing brickwork features. Undertake some individual research to:

- identify the shapes of some special bricks
- produce a small sketch of them.

Mortar

The materials used to make mortar are: sand, cement, lime and water.

Sand

The colour of sand can be altered by the addition a chemical colourant. This treats the sand grains and coats them in a colour. When the sand is mixed with water and cement it produces a coloured mortar. This can be used to alter the appearance of the joints and add a further architectural detail to facing brickwork.

Cross reference

Look at Unit 11, pages 119–120 for more information about sand, cement, lime and water.

Lime

Lime is added to mortar if the brickwork forms part of a restoration project. Lime was traditionally added to improve the workability of mortar in laying bricks. It is used in mortar in conservation work to replicate the methods and materials used in the past.

Assessment activity 13.2

You are training to be a bricklayer and have been asked to order in the materials for the following brickwork tasks:

- an isolated pier in English bond
- construction of a manhole.

Make a list of the materials that you would need. **P3**

Grading tip

To achieve **P3** you need to think about the two different areas where the materials will be used for each and identify these.

PLTS

Reflective learner think where these skills and techniques might be used

Functional skills

English write your list of materials by hand, neatly and clearly

2. Health, safety and welfare issues for brickwork tasks

2.1 Health, safety and welfare issues

Cross reference

Look at Unit 2 pages 31–40 for more information about risk assessment and controlling hazards.

Health, safety and welfare issues for the construction industry as a whole are covered in Unit 2 of this book. Look back at this unit to remind yourself of these. You should also refer to Unit 11 (pages 121–125), which covers health and safety issues for brickwork and blockwork, including the correct types and use of PPE.

Maintenance of a clean and tidy work space

When starting bricklaying, it is important that your materials are stacked correctly and that you use a mortarboard for the mortar. Any cut bricks should be placed in a wheelbarrow and disposed of into a skip, so they can be used later for crushed hardcore. A clean and tidy work area reduces the likelihood of an accident from a slip, trip or fall.

Identification of hazards associated with given tasks

Bricklaying has a number of construction site hazards associated with it.

- Bricks are packed and often shrink-wrapped or banded. If the packaging breaks or is damaged, a pack can collapse when it is lifted and the bricks can spill out.

- Bricks are smaller than blocks, and take more manual handling to stack them ready for use by the bricklayer.

- Cutting bricks with a brick hammer can lead to crush injuries.

- When bricks are being trimmed, sharp particles can fly off.

You should undertake a simple risk assessment for any construction activity to identify any hazards associated with the work.

Use of PPE to minimise risks from identified hazards

Bricklayers must wear a hard hat. Their job often involves laying bricks at height, which requires extensive use of scaffolding. A bricklayer will have to walk under the scaffolding, climb ladders up the scaffolding, and bend through guard rails. Each of these activities puts bricklayers at risk of banging their head, causing injury.

A second hazard is the many materials stacked on scaffolds. These have the potential to be knocked off and cause a head injury.

2.2 Hazards

The photo shows a building site for a typical home constructed of brickwork. There are many potential hazards associated with work on this type of site. The main hazard is the use of the scaffolding, as falling from height is the main cause of fatalities.

Slips, trips and falls

There are many trip hazards that people performing brickwork operations need to be aware of. Care should be taken to keep the work area free from the following types of hazard:

- trailing electrical leads
- ladders laid on the ground
- loose materials from laying bricks.

What hazards can you spot?

Case study: Bricklayer wins a million

The **UCAT** solicitors O.H. Parsons successfully won a High Court case for a bricklayer who sustained a severe fracture to his vertebrae, leaving him mainly wheelchair-bound, after the scaffold platform collapsed. He was building a small double-skin wall on the third lift of a scaffold and he fell through the scaffold boards about six metres to the lower platform. Court proceedings went all the way to a hearing against his employers, the scaffolding subcontractors and the main contractors.

Did you know?

In 2007/8 there were 34 deaths and 2000 fractured bones from falls from height.

Key term

UCAT – Union for Construction and Allied Trades.

Cuts and injuries caused by sharp tools and instruments

Great care must be taken with brick-cutting tools. A masonry saw can cause injury through:

- incorrect use
- not using guards correctly
- dust and flying particles
- noise from constant use.

You must have training before using rotating saws, and complete an abrasive wheel course before changing the blade, as this has to be done correctly.

Musculoskeletal injuries resulting from lifting and moving heavy loads

Because bricks are smaller than blocks, the risk from lifting or manual handling is lower. However, to transport bricks along a scaffold, they must be placed in a barrow, which then increases the weight that must be moved in one load.

In what ways can use of a brick saw's rotating blade cause injury?

 Cross reference

Look at Unit 8, page 73, for more details of how to lift correctly.

2.3 PPE

When performing brickwork operations you should always wear:

- safety boots
- hard hat
- high visibility jacket
- gloves
- overalls.

BTEC Assessment activity 13.3 P5 P6 M2

You are working on a project creating the garden at the back of a new house. This involves two brickwork tasks:

- building an isolated pier in English bond
- building a garden wall one brick thick in half bond.

1 Identify the PPE and safe working practices used to perform the two tasks. **P5**

2 Explain why you selected these PPE and the safe working practices for the two brickwork tasks. **P6**

3 Give good reasons for your choice of appropriate PPE and safe working practices to minimise health, safety and welfare risks for the two tasks. **M2**

Grading tips

1 To achieve **P5** you must look at the operations that are going to be performed and see if they will involve cutting bricks, which will involve specific PPE.

2 To achieve **P6** you must clearly outline your reasons for selecting each particular item of PPE.

3 To achieve **M2** you need to expand on your reasons to explain why you have selected each item of PPE: for example, would you need a hard hat outside?

What items of PPE is this bricklayer wearing?

3. Safe working practices to set out and construct brickwork

3.1 Stages of setting out

Correct calculation of lengths of walling detail

Dimensions for buildings are given by the architect or designer. They should have taken into account the modular dimensions of the bricks that are going to be used. Wall lengths will need setting out to full bricks, leaving a half-brick cut where possible. This makes the wall look right when it is finished. You may need to make adjustments within the wall by closing down the width of the mortar joint: for example, if you take 1 mm off 20 joints, you will give yourself 20 mm to play with to work the wall into stretchers.

Correct bonding of walling

The bonding of a wall will depend on its function: for example, an external wall to a house will have to be built in half bond as no other bonding can be used. Half bond just means that each brick is a half brick in front of its neighbour.

Garden walls or one-brick walls can be built using different bonds, such as English and Flemish bonds.

Half-brick walling

Corners in stretcher bond or half bond

Fig. 13.1 illustrates how the different courses alternate for a stretcher bond corner half brick thick. Note how each course overlaps the other by half a brick.

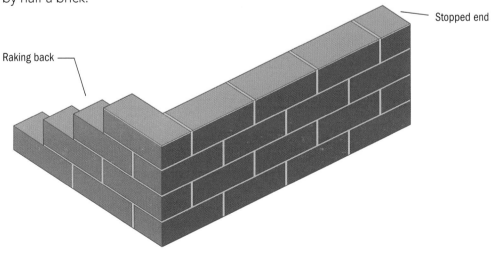

Raking back

Stopped end

Fig. 13.1: Stretcher bond corner

One-brick walling

Corners in English bond and Flemish bond

Fig. 13.2 and 13.3 illustrate the bonding arrangements for the two different corners. This will help you when setting out your work and constructing the corner correctly.

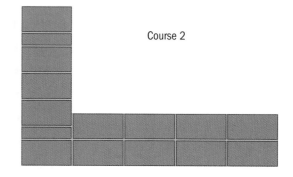

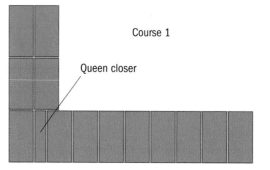

Fig. 13.2: Alternate courses in an English bond corner

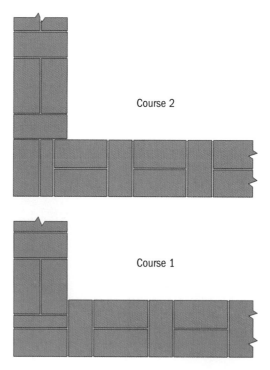

Fig. 13.3: Alternate courses in a Flemish bond corner

Attached piers in English bond and Flemish bond

These are difficult models to build and set out. Take great care with the setting out and draw it out on paper first.

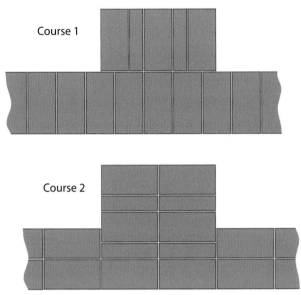

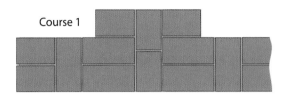

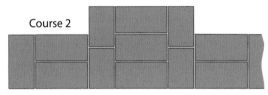

Fig. 13.4: Alternate courses of an attached pier in English bond

Fig. 13.5: Alternate courses of an attached pier in Flemish bond

Two brick walling

Isolated piers in English bond and Flemish bond

Fig. 13.6 shows the alternative courses in an isolated pier in English bond. These are difficult models to build and construct at this level and will challenge you.

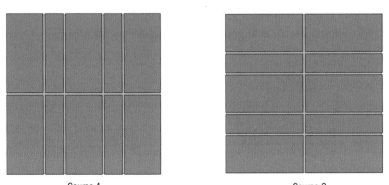

Fig. 13.6: Alternate courses of an isolated pier in English bond

BTEC Assessment activity 13.4 P2 P4 P7 P8 P9 M1 M3 D1

As part of your training you have been entered into a regional brickwork competition. You have to build the following models:

- A half-bond corner, ten courses high in stretcher bond.
- A one-brick English bond wall.

1 Select the hand tools you would need to perform these given brickwork tasks. **P2**

2 Select the materials required to perform these tasks. **P4**

3 Draw out the correct bonding arrangements for each alternative course on paper to scale. **P7**

4 Set out and produce the two brickwork models. **P8** **P9**

5 Justify your use of hand tools and selected materials to minimise health, safety and welfare risks. **M1**

6 Produce finished work with bricks in line to ±3mm, face plane deviation ≤3mm and corners plumb to within ±3mm per metre height. **M3**

7 Produce finished work with bricks in line to ±3mm, face plane deviation ≤3 mm and corners plumb to within ±3mm per metre height. **D1**

Grading tips

1 To achieve **P2** you will need to correctly select the hand tools for this operation and have this witnessed on an observation record by your tutor.

2 As with **P2** to achieve **P4** you need to select the right material, recording this on an appropriate document that has a witness signature.

3 To achieve **P7** you must be able to show that you understand what the different bonding arrangements are like, describing each clearly.

4 To achieve **P8** and **P9** you must produce the practical elements, remembering to take photographs of your work. Here you can get a friend to take your picture while you are constructing the models, and vice versa.

5 To achieve **M1** you need to expand on your earlier answers and give reasons why you have chosen the tools and materials you have used.

6 To achieve **M3** and **D1** you must make sure that your wall exactly fits the dimensions stated. Practice will improve the quality of your work.

James Nash

Bricklayer

James has been a qualified bricklayer for a year, after his NVQ Level 3 and apprenticeship. He began his training at school. His school was partnered with a local college and delivered the BTEC First Construction to Years 10 and 11.

James really enjoyed this course and continued at the local college, taking a NVQ Brickwork Level 2. He obtained a credit award in the end phase test, which his tutors were especially pleased with.

While at college James entered several national brickwork competitions and came second in a number of these. This has got James noticed not only by the college but also by his employer.

When James left college, he took a job as bricklayer with a national construction firm. They were very impressed by his experience, and the evidence of his self-motivation and commitment to the job. James worked really hard at his bricklaying and won the 'Apprentice of the Year' competition in his first year with the company. They are really pleased with his progress so far and he has featured in the company newsletter. After a year of work, he now wants to go back to college to continue to develop his skills on conservation brickwork.

James would like to learn how to construct complex masonry structures and produce decorative brickwork. This often involves stonework and he wants to try to develop skills in the replacement of this work as the company has a conservation arm. He even thinks that one day he could specialise in working on historic listed building restoration projects. Learning complex skills would make him a highly valued tradesman. James needs to find a course in this field which is very specialist.

Think about it!

1 What would you recommend to James?
2 What information about himself, his studies and career should he include on his college application?
3 What could you do outside your course to develop your brickwork skills?

Just checking

1 What other name is half bond known by?
2 Draw the alternate courses of brickwork for a half-bond stretcher corner so a colleague could build it.
3 Is there more than one method of pointing brickwork?
4 What is the difference between English bond and Flemish bond?
5 What hazards are associated with training mortar?
6 What does LBC stand for?
7 What is a frog in a brick?
8 How do corner blocks fix to brickwork?
9 Is an engineering brick heavier than a LBC?
10 Where are engineering bricks best used?

edexcel

Assignment tips

- Brickwork has to be set out correctly. Your eye will immediately notice that something is not quite right when it is looked over. Bonding is all about making the wall look great. Exercising great care in setting out the brickwork will only add to the quality of the finished work.

- Take time to keep the brickwork clean and tidy as this makes the work very neat and produces a quality finish.

14 Exploring painting and decorating

Most people think of painting and decorating as being about making a surface look attractive. After all, from the time of early cavemen, people in every culture have used pigments and colours to decorate their homes. However, there are other reasons to apply paint, including providing protection for a surface, to colour-code items or areas, and to produce surfaces that are easy to clean.

In this unit, you will discover the tools, equipment and techniques used by painters and decorators today, the different surface preparation materials and finishes that the modern decorator can use. As well as developing your knowledge and understanding of painting and decorating, you will carry out practical activities to develop skills in preparing surfaces, using tools, using equipment and applying paints. You will also learn about the importance of safe working practices.

Demand for decorating skills is greater than ever, and this unit will help you develop skills that are very much in demand, as well as preparing you for other related units in this qualification, such as paperhanging, applying textured finishes and installing coving and centrepieces.

Learning outcomes

After completing this unit, you should:

1 know the hand tools, materials and access equipment used by decorators to perform specified tasks

2 understand safe working practices to prepare new and previously painted surfaces for painting

3 be able to apply safe working practices in the application of paints to prepared surfaces.

Assessment and grading criteria

This table shows you what you must do in order to achieve a pass, merit or distinction grade, and where you can find activities in this book to help you.

To achieve a **pass** grade the evidence must show that you are able to:	To achieve a **merit** grade the evidence must show that, in addition to the pass criteria, you are able to:	To achieve a **distinction** grade the evidence must show that, in addition to the pass and merit criteria, you are able to:
P1 identify hand tools and access equipment used to perform painting and decorating tasks **See Assessment activity 14.1, page 168**	**M1** explain the safe use of tools, materials and access equipment to minimise health, safety and welfare risks **See Assessment activity 14.1, page 168**	
P2 select hand tools and access equipment used to perform painting and decorating tasks **See Assessment activity 14.1, page 168**		
P3 identify materials used to perform specified tasks **See Assessment activity 14.2, page 171**		
P4 select materials used to perform specified tasks **See Assessment activity 14.2, page 171**		
P5 identify the PPE and safe working practices used to perform specified tasks **See Assessment activity 14.3, page 176**	**M2** justify the use of appropriate PPE and safe working practices to minimise health, safety and welfare risks **See Assessment activity 14.3, page 176**	
P6 explain the selection of PPE and safe working practices used to perform specified tasks **See Assessment activity 14.3, page 176**		
P7 follow manufacturers' guidelines when preparing materials for use **See Assessment activity 14.4, page 176**	**M3** produce finished work with surfaces abraded without scoring and filled smooth, and finished paintwork bristle-free with all cutting-in sharp and neat **See Assessment activity 14.4, page 176**	**D1** produce finished work with surfaces abraded without scoring and filled smooth, finished paintwork bristle-free with all cutting-in sharp and neat and with no misses or other impairments evident from 1 m **See Assessment activity 14.4, page 176**
P8 perform painting and decorating activities using hand tools **See Assessment activity 14.4, page 176**		
P9 demonstrate the safe use of materials when performing painting and decorating tasks **See Assessment activity 14.4, page 176**		
P10 demonstrate the safe use of low level access equipment when performing painting and decorating tasks **See Assessment activity 14.4, page 176**		

How you will be assessed

This unit will be assessed by an internal assignment that will be designed and marked by the staff at your centre. Your assessment could be in the form of:

- presentations
- case studies
- practical tasks
- written assignments.

Richard, 16 years old

Painting and decorating is such a great way to transform a room. There are so many colours and coverings to choose, and so many different effects you can get. However, this unit showed me that there's a lot to learn if you want to make painting and decorating more than just a hobby.

For this unit, I had the chance to find out about all the different reasons people have for decorating their homes, and the other places we all use. It's amazing just how many different tools and materials there are, and all the equipment there is these days to help you. It's not just about a brush and a splash of paint!

The bit I enjoyed most was learning about the different paint finishes and surfaces you can create, because it's the creative side that appeals to me most.

Being able to put into practice what I'd learned was really useful too. I had the chance to use water-based and solvent-based paints, on different surfaces, and actually see the difference for myself. Now I feel I could really help a customer get the results they are looking for.

Over to you

- What do you think will be the most difficult parts of this unit?
- Which areas do you think will interest you the most?
- How much practice do you think you will need to prepare for the assessments in this unit?

1. Hand tools, materials and access equipment used by decorators

Not just a pretty face

Did you know there are four reasons for painting? Try to think of the four reasons why a surface is painted. You should think about the different locations where paint is used and what purposes it serves. Once you have identified the four reasons, write down an example for each.

1.1 Hand tools

Decorating encompasses a wide range of different activities, all of which require specific tools and equipment. These activities can be broadly classified as preparing surfaces and applying surface finishes.

Surface preparation tools

All surfaces must be properly prepared before painting or applying other finishes, whether they are new or previously painted. In most cases, new or previously unpainted, surfaces require relatively little preparation before painting. This could be something as simple as a light rub down and the removal of dirt, dust or grease. However, previously painted surfaces generally require much more preparation, depending on whether they are classed as **sound** or **unsound** surfaces. Sound surfaces may have some areas of paint film breakdown, such as flaking, cracking or blistering, but these will usually be quite small. Such areas should be scraped off and the bare surface underneath spot-primed. Unsound surfaces should be completely stripped, cleaned and primed before you apply any further finishes.

The range of tools used for preparing surfaces includes scrapers, filling knives, putty knives, shavehooks, caulking boards, hacking knives and hot air strippers.

Key terms

Sound surface – a surface that has previously been painted and is in generally good condition.

Unsound surface – a surface on which the paint film has broken down severely over most of its area.

Tang – the section of metal inside the handle of a scraper or knife.

Scraper

A scraper is used for scraping away loose or flaking paint or for removing old wall coverings. It has a stiff blade and **tang** made from one piece of metal with a hardwood handle.

Scraper

Filling knife

A filling knife is used for applying filler to cracks and other surface imperfections. It is very similar in construction to a scraper but with a more flexible blade.

Filling knife

Putty knife

A putty knife is used for smoothing putty around the edges of the glass when glazing. It is similar in construction to scraper and filling knife but it has a narrow blade that is straight on one side and curved on the other.

Putty knife

Shavehook

A shavehook is used for scraping paint from mouldings and other curved or shaped areas. It is normally used in conjunction with liquid paint remover or when burning off. There are three different shapes: triangular, pear-shaped and combination.

Shavehook

Caulking board

The flexible blade of a caulking board is used to apply filler or jointing material. It is usually made of a broad piece of plastic with a wooden handle.

Caulking board

Hacking knife

A hacking knife is used with a hammer to remove or hack out old putties when removing glass from window frames. The heavy metal handle has a flat top edge so it can be struck with a hammer. The knife has a leather handle to absorb impact and reduce vibration to the user.

Hacking knife

Hot air stripper

This electrical appliance is used to **burn off** old paint films. The stripper produces hot air at temperatures up to 600°C. You should only use a hot air stripper after proper training and when your tutor says you are competent to do so.

Hot air stripper

Tools for applying surface finishes

Tools for applying surface finishes include paint brushes, rollers, paint kettles, trays, and scuttles. They have different properties and uses.

Paint brushes

Paint brushes are used to apply paint to a surface. Different sizes are used depending on the size of the area being painted. They are available in a range of sizes from 12 mm to 100 mm and come with a number of different fillings for use with different paint types. Pure bristle brushes are best for solvent-based paints. Brushes are also used for cutting in when applying paint to large areas by roller.

Paint brush

Paint rollers

A roller is used to apply paint to broad flat or textured surfaces. It is generally much quicker to use a roller than a brush to cover large areas. They are available in a range of different widths and with different **sleeve** coverings and **pile lengths**. The frames which hold the roller are available in either single or double arm types.

Paint kettle

A kettle is used to hold paint when using a brush. Paint kettles are made of either plastic or galvanised steel.

Paint kettle

Roller tray

Made of either plastic or metal, a roller tray holds paint when using a roller.

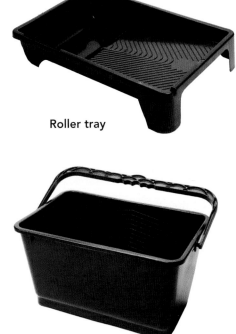

Roller tray

Scuttle

A scuttle can hold more paint than a roller tray. Some scuttles have hooks on the back so they can be hung on the rung of a ladder when you are working at height, although using a scaffold or other working platform is recommended.

Scuttle

Cleaning and maintaining tools

Decorators rely on their tools to produce high-quality finishes. Because of this, it is important that tools are properly cleaned and maintained.

Paintbrushes and rollers used with water-based paints and finishes should be cleaned using warm water. It is important however, that you do not use water that is too hot, as this can cause the paint to solidify in the brush and make it harder to clean. Once you have removed all of the paint, you should shake brushes to remove excess water and place rollers back on their frames, then spin them and hang them to dry.

When brushes have been used in solvent-based paints, you should clean them using turpentine substitute or white spirit to remove all the paint, rinse them with warm, soapy water to remove all of the solvent, then shake them to remove any excess water.

However, if you are going to reuse the brush in the same colour, there are alternatives to cleaning the brush after each use. For example, brushes can be stored in a galvanised steel box containing a small amount of special solvent, which releases a vapour that prevents the paint from drying in the brush. As well as saving time, this is a much more environmentally friendly alternative to cleaning brushes with solvents, and reduces environmental waste costs.

Kettles, trays and scuttles used with water-based paints should be cleaned using water. However, if you are using them with solvent-based paints, it is normal practice to brush excess paint back into the tin and then leave the kettle, tray or scuttle to dry before using it again.

Did you know?

Water-based paints dry faster than solvent-based paints.

BTEC Assessment activity 14.1

You are working as a decorator, and have been invited to a school to show the children the different sorts of tools that you use in your work. The class teacher has asked you to do two things:

1 explain what the tools are (including access equipment)

2 say what jobs they are used for.

You realise that you should also explain the safety issues around using these tools, so that the children can explain safe working practices to their parents!

3 Create a list of the different aspects of health and safety you would need to talk about, and say what the children's parents would need to do.

Grading tips

1 To achieve **P1** you will need to name each of the different tools involved in painting and decorating work. You could do this using pictures, with a sentence to name each and say what it does.

2 To achieve **P2** you will need to match the different tools to different painting and decorating tasks. You could add notes to the pictures above to say the sort of work each is used for.

3 To achieve **M1** you will need to explain clearly what the safety issues are concerning the different tools and equipment. You need only cover access equipment to reach a ceiling.

1.2 Materials

As with tools, the materials that a decorator uses can be classified as preparation materials and finishing materials. As you have seen, all surfaces must be properly prepared before surface finishes can be applied. The existing condition of the surface will determine the extent of the preparation. Preparation tasks include the complete removal of old paint systems using heat or solvents, scraping off loose or flaking finishes, filling holes, cracks or other surface imperfections, washing dirt or grease off a surface, or simply rubbing it down to make it smooth.

Surface preparation materials

Surface preparation materials are used to make surfaces ready for painting. Preparation materials are used to ensure a surface is clean, smooth and ready to receive a new coat of paint.

Liquid paint removers

Old paint systems that have broken down can be completely removed using liquid paint removers. These are spirit-based and work by breaking down the paint film and causing it to blister and lift from the surface. Most liquid paint removers require the surface to be neutralised after use by washing with water. However, there are some paint removers available that are **self-neutralising**.

Washing materials

Some surfaces require washing to remove dirt or grease before painting. This can be done using various detergents and washing materials. The most popular material for this is sugar soap.

Abrasives

Most surfaces require some form of abrasion, or rubbing down, before painting. A surface can be abraded using abrasive papers, abrasive powders or compounds or steel wool. The most common abrasive used by the decorator is aluminium oxide paper, which is available in a wide range of grades.

Abrasive papers are known as coated abrasives. The size of the abrasive particles determines the grade of the paper: for example, a paper graded as 60 has relatively large abrasive particles and is best suited to preparing surfaces in poor condition, while a paper graded as 400 has much smaller particles, and is therefore a much finer abrasive, ideal for preparing an undercoated surface before producing a high-quality gloss finish.

Some coated abrasives, such as silicon carbide, have a waterproof backing paper and can be used with water to eliminate dust; other coated abrasives such as aluminium oxide have a non-waterproof backing and can only be used dry.

Fillers

Fillers are available in either powder or ready-mixed forms. The choice of filler is determined by a range of factors including the surface being filled, the required drying times, the extent to which the surface is damaged, and the quality of the required finish. For example, where a surface has only minor imperfections, you can use ready-mixed fine surface filler; however, where a surface such as plaster has major cracking or other defects, you will need to use a powder-based filler.

Key term

Self-neutralising – for a paint remover, when the liquid neutralises itself, so the surface you use it on does not need further washing afterwards.

Did you know?

When carrying out decorating tasks, always keep a piece of abrasive paper, a scraper and a dust brush in your overall pockets.

169

Surface finishing materials

Paints are available in a wide range of different finishes and colours. Some paints are water-based, thinned with water, and some are solvent-based, thinned with white spirit. Table 14.1 shows some of the most commonly used paints and their uses.

Table 14.1: Surface finishing materials, their uses and properties

Key term

Opacity – the quality of lacking transparency, so that you cannot see through it.

Name	Water-based	Solvent-based	Use	Properties
Undercoat	Yes	Yes	Used before finish coat to build film thickness and cover previous coatings	Intermediate coat that has high opacity to help cover previous paint systems
Primer	Yes	Yes	The first coat of paint that is applied to a bare surface	Special primers are required for different types of surface: wood primers, primers for ferrous metals and non-ferrous metals, primers for plaster, etc.
Emulsion	Yes	No	Internal walls and ceilings	Available in matt, semi-matt or silk finish. A quick-drying, washable finish available in a wide range of colours
Eggshell	Yes	Yes	Mainly used on internal woodwork. Can also be used on walls and ceilings	Available as water- or solvent-based. Eggshell dries to a semi-gloss finish that can be wiped clean
Gloss	Yes	Yes	Mainly used on internal and external woodwork. Can also be used on walls and ceilings	Dries to a hard gloss finish that can be easily wiped clean. Very good weather resistance makes it an ideal finish for exterior surfaces like doors and windows

Assessment activity 14.2 · BTEC · P3 P4

Your boss is preparing to quote for a job, and has asked you to prepare a specification for the preparation and finishing of the two surfaces described below.

1 An exterior softwood door that has large areas where the paint film is badly breaking down and flaking.

2 An internal wall that is finished in vinyl matt emulsion. The paint film is in generally good condition but is a bit dirty. The surface has a small number of cracks and other minor imperfections

Grading tips

1 To achieve **P3** you should name all of the materials required to properly prepare the softwood door and to prepare the wall area for re-painting.

2 To achieve **P4** you should include all of the appropriate materials in your specification, and not include any that are not needed.

PLTS

Team worker discuss possible specification options with others

Functional skills

English use appropriate language in your specification

1.3 PPE

The decorator often works with tools and materials that can be harmful, especially solvent-based materials that can irritate the skin. When working with such materials it is important that you protect your skin, using either gloves or barrier cream. You must remember that, before eating or drinking, you need to remove all traces of paint, solvent or barrier cream from your hands to prevent **ingestion**.

In addition to the risks presented to the skin, there are also issues around personal cleanliness while performing surface preparation and painting tasks. You should always wear overalls while decorating: this will help to keep your clothes clean and prevent dust and paint being transported away from the work area. To prevent potential injury while working with heavy items (for example, when using ladders or stepladders), you should also wear safety footwear.

Another potential risk to the decorator is that of eye injuries. These most commonly occur when working on ceilings internally, or when preparing and painting fascia boards, gutters and soffits externally. This type of work often involves looking up while you work, and this is when small particles of old or new paint, dust or other debris can easily get into your eyes if they are not protected. Usually, safety glasses are sufficient to prevent this type of injury.

Key terms

Ingestion – taking into the body by eating or swallowing.

Cross reference

Look at Unit 2 for more information about PPE.

1.4 Access equipment

The decorator often has to work at height. This may be working only a few inches from the ground in order to reach the top of a wall or a ceiling when working internally, or at greater heights when working externally, in order to gain access to first-floor windows, gutters, fascia boards and soffits. Working at height requires the use of a range of access equipment. When working internally, low-level access equipment is usually sufficient. Items of low-level access equipment include stepladders, hop-ups and trestles used with lightweight staging.

Stepladders

Stepladders are available in a range of different sizes for a wide range of different tasks. They are used both internally and externally to access work above ground level. The stepladder is the piece of access equipment most commonly used by decorators. It is important to ensure that stepladders are kept in good condition and are always fully opened and on a firm, level base when in use.

Hop-ups

These are used to access work just above normal reach, such as a ceiling in a modern house, which is normally around 2.4 metres high. As with stepladders, they must be kept in good condition and used on a firm level base. A typical hop-up will have a working platform of around 500 mm × 400 mm and a height of around 400–500 mm.

Trestles and staging

Trestles and lightweight staging are normally used when you need to access to a large area or long run of work. Trestles are made from either timber or aluminium, and are formed using an A-shaped frame. As with stepladders, trestles must be fully opened when in use and must be on a firm, level base. The trestles are used to support lightweight staging, which can be made from timber, aluminium or a combination of both.

When using access equipment it is important to ensure that you follow safe working practices at all times. This includes making sure that all equipment is safe and in good condition before use, that it is kept clean and that any damaged or faulty equipment is not used.

What safety precautions should you take when using access equipment such as a stepladder?

2. Safe working practices to prepare new and previously painted surfaces for painting

2.1 New and previously painted surfaces

New surfaces generally require much less preparation than those that have been previously painted. However, it is important to ensure that appropriate preparation is carried out before applying paint to any surface, whether it is new or not. Table 14.2 below gives some examples of typical preparation activities for a range of new and previously painted surfaces.

Did you know?

Using water-based paints can help to reduce the build-up of fumes in confined spaces.

Table 14.2: Preparing new and previously painted surfaces

Surface	Condition	Preparation
Timber	New and unpainted	Lightly rub down to ensure the surface is clean and smooth, apply two coats of shellac knotting to prevent resin leaking from knots and apply solvent or water-based wood primer.
	Previously painted and in sound condition	Rub down to ensure surface is clean, smooth and free from any loose or flaking materials. The surface may also need to be washed to remove dirt or grease before applying solvent or water-based undercoat.
	Previously painted and in unsound condition	Remove all traces of previous coating by burning off or using liquid paint remover. The surface can then be treated as new, unpainted timber.
Ferrous metals	New	Clean with white spirit to remove all traces of grease before applying appropriate primer.
Non-ferrous metals	New	Clean with white spirit to remove all traces of grease before applying appropriate primer.
Plaster	New, to be painted with water-based paint system	Lightly rub down to ensure surface is clean and smooth before applying water-based primer or first coat of emulsion.
	New, to be painted with solvent-based paint system	Lightly rub down to ensure surface is clean and smooth before applying alkali-resisting primer.
Plasterboard	Previously painted and in sound condition	Lightly rub down to ensure surface is clean and smooth with no loose or flaking material. Fill surface imperfections and rub down again before painting.
Masonry	Previously painted and in sound condition	Jet wash or scrub with a suitable detergent to ensure surface is clean. Allow to dry thoroughly before applying two coats of masonry paint.

Activity: Restaurant revamp

Harriet is a self-employed decorator employing two other decorators and a second-year apprentice decorator. She has been asked to provide a specification proposal and quote for the external redecoration of a medium-sized, two-storey family restaurant with flat above. The restaurant owner has indicated that he will ask her to quote for the internal redecoration as well if he is happy with the standard of Harriet's work.

The restaurant is in the town centre, next to a busy main road. Consequently, the exterior paintwork gets very dirty. It was last painted seven years ago and many of the windows are in an unsound condition. The gutters and fascia boards are plastic and do not require painting, but are very dirty. The front door to the restaurant is also very dirty but in generally sound condition.

- **What is the first thing that Harriet must do to help her quote for the job?**
- **What things will she need to consider when preparing the specification?**
- **What access equipment will Harriet need to complete the job?**
- **What would be the best finish to recommend for the windows and doors?**

3. Applying safe working practices in the application of paints to prepared surfaces

Why should you take care when working with paint?

3.1 Working with paints

As already discussed, paints can be broadly categorised as water-based and solvent-based. Safe working practices, while in some ways very similar for both, can differ depending on the type of paint and the location in which it is being applied. For example, applying emulsion paint to the walls of a large, well-ventilated room presents very few problems in terms of the build up of harmful fumes. However, applying solvent-based gloss finish to the woodwork in a small cupboard, or other confined space, requires the use of a respirator to prevent the inhalation of fumes. Similarly, the use of water-based paints generally presents fewer risks to the skin than using solvent-based paints. However, the decorator should always take every precaution to minimise the amount of any paint that comes into direct contact with the skin. The use of barrier cream or gloves can help to ensure this. Remember though, any paint, barrier cream or other material should always be removed from the hands before eating or drinking.

3.2 Health, safety and welfare

While at work, every individual is responsible for ensuring the health, safety and welfare of themselves and others around them. There are a number of ways in which the decorator can ensure that they do this.

Maintaining a clean and tidy work space

Ensuring that your work area is kept clean and tidy can help to reduce the risk of accident or injury. Picking up tools, materials and equipment that are not being used and storing them out of the way can reduce the risk of tripping. Similarly, cleaning up spillages immediately can reduce the risk of slipping. Another benefit is the reduced likelihood of dirt or debris getting onto a freshly painted surface.

Identification of hazards associated with given tasks

Identifying hazards, risks and control measures should always be carried out before starting a job or task. In some situations a good 'risk assessment' and the use of control measures can completely remove the risk of accident or injury. However, there are times when these risks cannot be completely removed. In such cases, control measures must be identified to minimise the risk of accident or injury.

COSHH

COSHH places responsibility with employers to take steps to prevent employees' health being harmed by substances used in the workplace. The COSHH regulations require risk assessments to be carried out when working with substances that are harmful to health. Employers must carry out the following to prevent or reduce employees' exposure to hazardous substances:

* finding out what the hazards are
* deciding how to prevent harm to health (risk assessment)
* providing control measures to reduce harm to health
* making sure control measures are used
* keeping all control measures in good working order
* providing information, instruction and training for employees and others
* providing monitoring and health surveillance in appropriate cases
* planning for emergencies.

The employee also has a responsibility to use the control measures identified when using materials that are hazardous to health.

Health, safety and welfare are paramount with decorating tasks. There are many risks and hazards associated with using paints, solvents, tools and equipment. Using risk assessments and identifying control measures can help to manage these (see Unit 2, pages 31–40).

PLTS

Independent enquirer take into account manufacturer's safety data sheets

Functional skills

ICT use the internet to research possible control measures

PLTS

Reflective learner think about how you might do this differently next time

Self-manager set yourself a realistic timescale and stick to it

Functional skills

Mathematics calculate the amount of paint you need in advance

---

BTEC Assessment activity 14.3 P5 P6 M2

You have been asked to create a safety poster that shows the range of hazards that the decorator faces while preparing surfaces and applying paints. Make a list of as many hazards and their associated risks and control measures as you can think of.

Grading tips

1 To achieve **P5** you must name and describe all of the hazards associated with surface preparation tasks, and identify the appropriate items of PPE and appropriate safe working practices for each hazard on your list.

2 To achieve **P6** you must explain why you have selected the various items of PPE and safe working practices. This will involve explaining how they work to reduce the hazards.

3 To achieve **M2** you will need to give reasons for your selections, based on the nature of the hazards and how the items and practices address these.

BTEC Assessment activity 14.4 P7 P8 P9 P10 M3 D1

The final assessment activity for this unit is the completion of a small painting task. You will be given a work area that consists of a ceiling and at least one wall, which you must paint to a given specification.

1 Prepare your ceiling area and apply two coats of white vinyl matt emulsion using both brush and roller. **P8 M3**

2 While painting your ceiling area, you will need to select and safely use appropriate low-level access equipment. **P10**

3 Prepare your wall area and apply two coats of acrylic eggshell finish in a colour of your choice. You must ensure that all cutting in to adjacent surfaces is sharp and neat, and that there are no misses evident from a distance of 1 metre. **P8 M3 D1**

4 You must follow all manufacturers' instructions while preparing materials for use and throughout their application. **P7 P9**

Grading tips

1 To achieve **P8** and **M3** you must complete all aspects of the tasks you have been given, and do so to an acceptable standard.

2 To achieve **P10** you must show that you know how to use access equipment safely.

3 To achieve **P7** you must make sure that you follow the manufacturers' instructions carefully as you prepare the materials for use. To achieve **P9** you must show that you can use the materials safely. To achieve **D1** you will need to take care that your finished work is of a good standard.

Sam Grove

Decorator

Sam completed a BTEC First in Construction during her last two years at school. She was particularly good at the more practical units, especially painting and decorating. This inspired Sam start an apprenticeship with a local decorating company immediately after leaving school.

Seven years later, Sam has completed her apprenticeship and has a further four years of experience with the same company. She is considered by her employer to be an excellent worker, and has been given increasing levels of responsibility and has run many small jobs on her own, taking a lead role when there has been a small team of decorators on site.

Sometimes she carries out a range of planning activities for jobs, and orders materials and equipment for delivery to site. This increasingly involves Sam being based in the company's offices for part of the working week.

Sam's employer is keen for her to develop this role further and to take on more responsibility for the planning, pricing and management of jobs. This would mean that Sam will be based in the office and will rarely work on site as a decorator. She has talked to Sam and asked her to identify her training needs for taking on this new role.

One possibility is to return to college part-time to complete the Level 3 BTEC National Award in Construction. This would broaden Sam's knowledge and enable her to specialise in areas like Planning, Organisation and Control of Resources and Measuring, Estimating and Tendering Processes, which would be particularly useful for the role that Sam's employer wants her to take on. It would also enable Sam to progress to a Higher National Certificate, which in turn could lead to a construction degree.

Think about it!

1 As well as meeting the needs of her current employer, how else will Sam benefit from returning to college and gaining additional qualifications?

2 How will Sam's site experience help her with her studies?

3 How will Sam's site experience help her in her new role?

Just checking

1 What is the main benefit of using hot air for burning off instead of a blowtorch?
2 What is the main difference between a scraper and a filling knife?
3 What is meant by an unsound surface?
4 When would the decorator use a hacking knife?
5 What would be a suitable alternative to using heat when stripping a surface of its previous paint system?
6 What is the range of widths that paint brushes come in?
7 Name two essential elements of a risk assessment.
8 How should the decorator prepare an unsound timber surface?
9 What should solvent-based gloss paint be thinned with?
10 State one advantage that a roller has over a brush.

edexcel :::

Assignment tips

- When you are out and about, look at surfaces that have been painted and think about why. Remember that there are only four reasons for painting.

- Make a visit to a trade centre and to your local DIY centre and look at the range of painting and decorating tools available. Look at the difference in quality between some of the tools you can buy and hire.

- Look around you and see how many different types of finish you can see. You can look at home, at school or college, at the shopping centre or anywhere else that you go.

15 Performing paperhanging operations

When wallpapering was first introduced to Europe, some 500 years ago, it was popular, but also expensive. With the introduction of machine printing it has become a relatively cheap method of decoration. Thanks to new washable materials and the wide variety of designs available, ranging from budget to designer styles it is 'trendy' again.

Wallpapers can add style, colour or texture to a room, but some are used to improve a surface before painting it, such as blown vinyl and Anaglypta™. Wallpaper can also be used to create different effects in a room, a vertical stripe will give the impression that a wall is taller.

In this unit you will learn about the different types of wallpaper, as well as the tools, equipment and techniques painters and decorators use to hang wallpaper. You will learn about the importance of safe working practices whenever you hang wallpaper.

You may study this unit as a stand-alone topic, or combine it with Unit 14 Exploring painting and decorating, and Unit 16 Performing decorating operations.

Learning outcomes

After completing this unit, you should:

1 know the hand tools, materials and access equipment used by decorators to perform paperhanging activities
2 understand the important health, safety and welfare issues associated with paperhanging tasks
3 be able to apply safe working practices when performing paperhanging tasks.

Assessment and grading criteria

This table shows you what you must do in order to achieve a pass, merit or distinction grade, and where you can find activities in this book to help you.

To achieve a **pass** grade the evidence must show that you are able to:	To achieve a **merit** grade the evidence must show that, in addition to the pass criteria, you are able to:	To achieve a **distinction** grade the evidence must show that, in addition to the pass and merit criteria, you are able to:
P1 identify the hand tools and access equipment used to perform paperhanging operations **See 185 activity 15.1, page 185**	**M1** justify the safe use of tools, materials and access equipment to minimise health, safety and welfare risks **See Assessment activity 15.1, page 185**	
P2 select the hand tools and access equipment used to perform specified paperhanging tasks **See Assessment activity 15.3, page 198**		
P3 identify the materials used in paperhanging operations **See Assessment activity 15.1, page 185**		
P4 select the materials used in specified paperhanging tasks **See Assessment activity 15.3, page 198**		
P5 identify the PPE and safe working practices used to perform specified paperhanging tasks **See Assessment activity 15.2, page 194**	**M2** justify the appropriate use of PPE and safe working practices to minimise health, safety and welfare risks **See Assessment activity 15.2, page 194**	
P6 explain the selection of PPE and safe working practices used to perform specified tasks **See Assessment activity 15.2, page 194**		
P7 follow manufacturers' guidelines when preparing materials for use **See Assessment activity 15.3, page 198**	**M3** produce finished work with no air bubbles, blisters or wrinkles and no gaps or overlaps >2 mm **See Assessment activity 15.3, page 198**	**D1** produce finished work with no air bubbles, blisters or wrinkles, no gaps or overlaps >2 mm, with straight, neat ends with no scissor marks, plumb deviation <2 mm from vertical and any patterns accurately matched **See Assessment activity 15.3, page 198**
P8 perform specified tasks using hand tools **See Assessment activity 15.3, page 198**		
P9 demonstrate the safe use of materials when performing specified tasks **See Assessment activity 15.3, page 198**		
P10 demonstrate the safe use of low level access equipment when performing specified tasks **See Assessment activity 15.3, page 198**		

How you will be assessed

This unit will be assessed by an internal assignment that will be designed and marked by the staff at your centre. Your assessment could be in the form of:

- presentations
- case studies
- practical tasks
- written assignments.

Laura, 15 years old

Going into this unit, I was worried that I would make a real mess of wallpaper hanging – I've seen comedy sketches of people trying to hang wallpaper, where they step in the bucket, and get paste and paper all over the place. I thought that might be what I'd end up doing! We did a role play that made me realise how serious the health and safety side of things is. It's so easy to get badly injured, and if that happens, it could end your career. By the end of the unit I could name all the tools you use for paperhanging, and I'd learnt all the tricks to use. That made me feel like a real professional.

This unit gave me the chance to see how versatile wallpaper is. It is pretty trendy now. I look around wherever I go – in modern houses and old homes, in restaurants and public places like the library – to see what wallcoverings they've used. Often I think 'I could do that', or at least, I will be able to soon. Maybe one day I could team up with an interior designer and offer a real high-end decorating service.

Over to you

- How can knowing about the tools involved in paperhanging help you?
- What skills and personal attributes do you think you need to be good at paperhanging?
- Are you considering a career in painting and decorating?

1. Hand tools, materials and access equipment

Build up

Everyone has their hang ups!

Did you know that the Chinese are believed to have glued rice paper to the walls of their homes as early as 200 BC? However, wallpaper as we know it today was first used in the early 1500s.

Where have you seen wallpaper? Why was it used here? Think about the type of place and the image it was trying to create, and note down your thoughts.

1.1 Hand tools

Paperhanging requires a different range of tools to painting, and most are designed for specific tasks. As with painting, surface preparation is vital to achieve a satisfactory end result.

Paperhanging brush

This brush with a hardwood handle and natural bristles is used to smooth down wallpaper onto a wall and remove air bubbles. The bristles are numerous and stiff enough to provide pressure during use.

Paperhanging brush

Paste brush or wall brush

A paste or wall brush is used to apply paste to wallpaper or a wall. Its long bristles speed up the application and it can also be used for **sizing**. It usually has a hardwood handle and copper **ferrule** and is available in sizes 100 mm to 175 mm.

Paste brush

Key terms

Sizing – Applying a special coating material to a surface to make it less absorbent.

Ferrule – A metal band that secures the bristles to the handle.

Seam roller

A seam roller is rolled down the joints when applying wallpapers. You must take care with the roller as overuse can cause sheen patches or **delamination**.

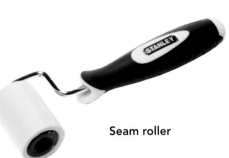

Seam roller

Caulk board

A caulk board has a flexible insert used to smooth out vinyl papers and remove air bubbles.

Caulk board

Paste table

Paste tables provide a long, flat surface to support wall coverings while cutting and pasting take place.

Paste table

Paperhanging shears

Shears are used to cut and trim paper while it is on the paste table or the wall. They have polished stainless steel blades with sharp cutting edges and must be kept clean to maintain a sharp edge.

Paperhanging shears

Tape measure

Tape measures are used to produce accurate measurements while measuring up and cutting paper.

Tape measure

Plumb line and bob

A string attached to a cylindrical weight (the bob) is held so it hangs against the wall. The string produces a perfectly vertical line, which can then be marked before starting to hang paper.

Plumb line and bob

Spirit level

A spirit level provides a vertical and horizontal line. It is ideal for use as a straight edge when producing a split paper. It is available in various lengths but 600–1000 mm is ideal for a decorator.

Spirit level

Trimming knife

A trimming knife is used with a straight edge to produce a neat cut at ceilings, door casings, etc. It has an extremely sharp blade that can be snapped off up to ten times using the slot at the opposite end.

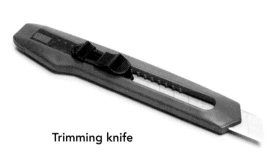

Trimming knife

Straight edge

This is used with a trimming knife to create straight cuts and prevent tearing. It is made from a bevel-edged length of steel.

Straight edge

BTEC Assessment activity 15.1 (P1) (P3) (M1)

You are working as part of a house-decorating team, and have been given an area of wall to hang paper on.

1   State what hand tools you will need when hanging wallpaper and what you will use each one for. (P1) (M1)

2   List the access equipment you will need when hanging wallpaper and state why you need each item. (P1) (M1)

3   List all of the materials that will be required to prepare a wall area, cross-line it and hang a vinyl wallpaper. (P3) (M1)

Grading tips

1   To achieve (P1) you should identify all of the hand tools and access equipment required to complete the task and outline clearly what each is used for.

2   To achieve (P3) you should identify the materials required to complete the given task and explain why you would choose each.

3   To achieve (M1) you should state why you have selected each hand tool, each material and each item of access equipment.

PLTS

Self-manager ask your tutor to help you make any changes needed

Team worker think about the needs of others working near you

Functional skills

ICT create a table showing tools needed for preparation, cross-lining and hanging

1.2 Materials

The only materials you need to hang paper are paste and wallpaper. However, there are many different categories for each of these. As with painting, the preparation for and choice of materials depend on the condition of the surface to which the paper is to be applied. Surfaces that you may paper include walls, ceilings, stairwells, and even arches and columns.

Porosity

Before papering, or even preparing, a surface, judging its **porosity** is of great importance. When a surface is porous, it is absorbent and soaks up moisture. High or uneven porosity can lead to defects such as blistering and lack of adhesion; uneven porosity also creates hot spots, which means that some patches will dry more rapidly than the rest. It will also mean that the paper is difficult to handle, causing lack of slip and poor pattern matching. This can usually be remedied by sizing the surface before hanging it.

To size a porous surface, apply thinned-down paste to it or use a proprietary glue size. Sizing the surface stops the moisture being taken out of the paste, preventing snatch, which is when the paper sticks to the surface without allowing any slip for pattern matching or manoeuvre.

Key term

Porosity – how porous or absorbent something is.

Why is it important to prepare surfaces carefully before papering?

Using lining paper provides a surface of even porosity, and creates the perfect base for applying finished papers, or even emulsion.

When a surface is non-porous, for instance if it has been painted with a gloss paint, this can cause just as many problems, especially with achieving adhesion. A non-porous surface will need to be **abraded** and must always be **cross-lined**.

Pastes

Although there are three main types of paste available, the modern decorator uses cellulose and ready-mixed pastes.

Table 15.1: Advantages and disadvantages of different adhesives

Type of paste	Advantages	Disadvantages
Cellulose	Does not stain the paper Inexpensive Contains fungicide Quick and easy to mix	High water content (97%) Lower adhesion than starch or ready-mixed
Starch	Good adhesion Less water content than cellulose	Does not contain fungicide, so should not be used for vinyl Needs to be used within two days once mixed Can stain the face of paper
Ready-mixed or tub paste	Contains PVA for a strong bond High solid, low water content No mixing involved	Expensive Harder to spread

Preparatory papers

Generally speaking, preparatory papers are lining papers. However the category can also include Anaglypta™, woodchip, blown vinyl, and speciality papers that deal with problem surfaces, such as damp.

Lining paper

This is a smooth wood-pulp paper, which makes an ideal base for finish papers. It is available in a range of grades, and in single, double and quad rolls. The standard width is 560 mm.

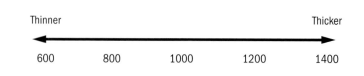

Fig. 15.1: Common grades or weights of lining paper in use

Lining paper provides a surface of even porosity, helps to hide minor surface defects and creates an ideal surface for painting. Using a lining paper on non-porous surfaces such as gloss or eggshell walls is crucial.

When hanging a lining paper, you must give it time to expand and become flexible.

If you are applying a finished paper over the top using a **butt joint** or a hairline gap, you should hang the lining paper horizontally. This prevents the corresponding joints from **springing** when the finished paper is applied. If the surface is to receive a painted finish then joints must be butted, and you should hang the lining paper vertically.

Anaglypta™

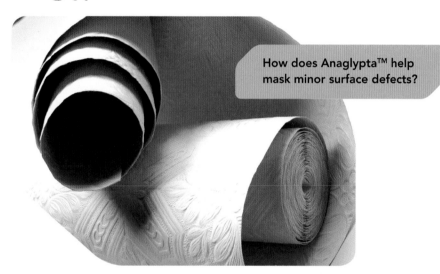

How does Anaglypta™ help mask minor surface defects?

This is an **embossed** wallcovering made from wood pulp. Once painted, it provides a hard-wearing surface.

Anaglypta™ which comes in a standard size of 10.05 m × 0.52/0.53 m is used to give walls and ceilings a textured look, and has the benefit of covering minor defects. It comes in a range of classic patterns as well as modern designs.

To hang Anaglypta™, apply a thick coat of starch or cellulose paste to the back of the paper. Take care not to over-soak it, as this can lead to crushing the pattern and delamination. Use a hanging brush to apply the paper, making sure you use an even pressure throughout the task; if you do not, it will result in flat patches of pattern. Anaglypta™ can be trimmed with shears or a knife and straight edge. However, you must take care when using a knife as the paper can tear easily.

Woodchip

Woodchip paper consists of two layers of paper with small wood pieces sandwiched in between. It is available in a standard size of 10.05 m × 520/530 mm, and gives a hardwearing surface once painted. It is a useful way of covering surfaces that are in poor condition.

Key term

Butt joint – when the edges of the paper are touching but not overlapping.

Springing – coming unstuck and sticking out.

Key terms

Embossed – with a raised pattern.

To hang woodchip, first apply a thick coat of cellulose or starch paste. There is no pattern to match, and the texture cannot be flattened, which makes it easy to hang. The woodchips can make it difficult to cut, so you should only use shears.

Blown vinyl

Blown vinyl paper consists of a paper backing with expanded PVC on the surface, which is used to form a random or repeating pattern.

With a standard size of 10.05 m × 0.52/0.53 m, blown vinyl paper offers another way to hide blemishes on walls and ceilings that are in poor condition.

To hang a blown vinyl paper, apply a stout cellulose or ready-mixed paste. You will find this sort of paper easy to paste, as it has a flat back. Use shears or a knife and straight edge to trim the paper.

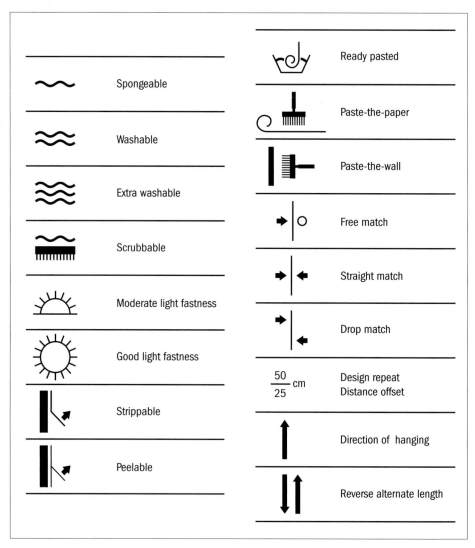

Fig. 15.2: International performance symbols

Finished papers

These are papers that require no further decoration, including washables and vinyls. They are available in a huge range of styles and patterns, from busy to plain.

Types of pattern

If a paper has a pattern, it will usually be either a **straight match** or a **drop match**.

Fig. 15.3: A straight match

Fig. 15.4: A drop match

Key term

Simplex – made of a single layer of paper, such as standard or washable paper.

Vinyl paper

With vinyl paper, a thin layer of smooth or textured PVC is bonded to a backing paper, producing a hard-wearing, washable surface. Vinyl paper comes in a standard size of 10.05 m × 520/530 mm, but contract widths of 1.3 m are also available.

You can use this sort of paper on most surfaces, including kitchens and bathrooms, and it is especially good for high-traffic areas, as it can be scrubbed clean if required and is resistant to stains.

To hang vinyl paper, you must first apply a paste containing fungicide. As it is a non-porous surface, the paste takes longer to dry out. You should cut it with a knife and straight edge, and can smooth it out with a brush or caulking tool.

Washable papers

Washable papers are usually **simplex** papers printed and then coated with PVA, to allow light wiping to clean. With a standard size of 10.05 m × 520/530 mm, washable papers are mainly used in domestic settings, as they have a high-quality look but are easy to keep clean.

To hang washable papers, paste with a cellulose or ready mixed adhesive then hang with a brush only, as a caulking board can damage the surface. You can use a knife or shears to cut this type of paper.

1.3 PPE

Although you are not using paint when paperhanging, you must still use personal protective equipment.

- Safety footwear should always be worn, and is a mandatory requirement on all site work.

- Overalls are worn to protect clothes from being contaminated by pastes. They should be cleaned regularly to stop the transfer of paste onto your hands and dirt onto the face of the paper. Wearing overalls also looks professional to the customer.

- Hand protection is sometimes used as some people can have an allergic reaction to paste, causing a rash on the skin. Hand protection can be in the form of barrier cream or latex gloves.

1.4 Access equipment

The type of equipment suitable for paperhanging includes stepladders, hop-ups and trestle working platforms. All of these let you access the work area without having to touch the walls, which means you can work efficiently, with less chance of damaging the wallpaper.

Stepladders provide easy access for the vast majority of paperhanging tasks in a domestic setting. Available in a range of sizes, from as few as two treads, they can be used on their own for papering walls, or two pairs with planks can be used for access when you want to apply lengths of paper to ceilings.

Hop-ups are good for low-level work, providing a height of around 400 mm and allowing freedom of movement when paperhanging. They are ideal for modern properties with lower ceilings.

A trestle working platform uses two trestles with a staging board spanning between them. This is classed as large access equipment, and would be used in large rooms with ceilings over three metres high. Trestle working platforms are a popular choice for commercial jobs, especially when using wide vinyl.

Why are stepladders often used in the domestic setting?

2. Health, safety and welfare issues associated with paperhanging tasks

When performing paperhanging activities it is important to ensure that your work area is kept as clean and tidy as possible. Tools, materials and equipment that are left lying around can create trip hazards and spilled materials such as adhesive can cause slip hazards. A clean and tidy work area also enables you to move yourself and your access equipment around more easily and safely. This also provides for more efficient working practices.

It is important to carry out a risk assessment before starting work. A good risk assessment will identify hazards, risks and who is at risk and suggest control measures to reduce the risk.

2.1 Pastes

Working with pastes should be safe, providing that you follow a few points.

- Always wash your hands after using paste and before eating, because most pastes contain fungicide, which can lead to an upset stomach.

- When working for customers, do not leave mixed paste where there is a risk of children or animals coming into contact with it.

- Be careful with trimming waste as this can cause slips. Always work tidily to avoid this. A tidy workspace always looks good to the customer too.

2.2 Techniques

Preparing surfaces

If a surface is to be re-papered, you will need to prepare it differently from a painted surface. You will need to remove old paste by washing the surface, which allows the removal of mould, should there be any. Note that you should never sand mould off as this puts spores into the air, which are a health hazard if inhaled.

Making good surfaces is the same as for painting, as you will need to fill all holes and cracks. The only difference is that the filler needs to be sized once it is dry, to even the porosity of the surface and make sure the lining adheres firmly.

Painted surfaces need to be rubbed down to **de-nib** them and provide a key. This requires a dust mask in addition to the normal PPE.

Bare plaster must be sized to reduce the porosity of the surface. Due to the thin consistency of size, the surrounding surfaces or areas need to be protected.

> ### Key term
>
> **De-nib** – lightly rub down to ensure a smooth surface.

Cutting wall coverings

Paperhanging operations are generally safe. However, because of the sharp tools needed for cutting, you should stay cautious by adhering to the following guidelines.

- Always retract the blade when not in use.
- When snapping blades off, use the slot on the opposite end of the knife.
- Put the used blades back in the plastic blade case.
- When washing shears, be careful with the blade edges as they can slice through cloth.

3. Safe working practices when performing paperhanging tasks

3.1 Access equipment

You always need to check your access equipment before use. You should check timber components to make sure that:

- there are no splits or cracks
- they have not been painted, because this can hide defects or repairs
- on steps and trestles, tie cords are of equal length and in good condition

- hinges are in good condition

- screws and bolts are all secure.

With aluminium components, you need to check that:

- there are no dents or damage

- rivets are all solid

- there is no corrosion.

If you follow the advice in Table 15.2, access equipment should be safe and reliable when in use.

Table 15.2: Using and storing access equipment

Equipment	Do	Do not	Storage
Stepladders	Always use fully open Ensure a firm, level base Ensure knees are below the top tread Use at 90° to the wall	Use as a ladder (closed) Stand on the top Use if treads are missing or broken	Closed together, standing upright Undercover/in a dry location
Hop-ups	Use on a firm, level base Ensure sturdy construction	Use something that is not designed to be a hop-up	Undercover/in a dry location
Trestles	Always use fully open Ensure a firm, level base Use with a staging board	Use as a ladder (closed) Use if treads are missing or broken	Undercover/in a dry location Closed with check blocks in place

3.2 Paperhanging tasks

Before you can start paperhanging, the job needs to be safely set up. Dustsheets should be placed as flat as possible to reduce any trip hazards and to protect the floor. You can then set up the paste table, which should be placed in the centre of the room if possible, to allow you to work around (and to save you having to move it about). Mix the paste using the manufacturer's information as a guide, making sure that you use the safety precautions we have already covered (see page 191).

Once a surface has been prepared and cross-lined, it is ready to receive the finished wallpaper. When papering a room, it is crucial to work away from the natural light, such as the window in a room, as this stops the joints casting a shadow. Once you have established your start point for, say, the window, the next step is to set up a plumb line 510 mm from the edge of the window (10 mm less than the width of the paper).

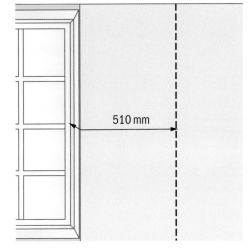

510 mm

Fig. 15.5: Establishing a starting point and marking a plumb line

Cutting the paper

You will need to cut the lengths of paper required by measuring the height of the wall from the top of the skirting to the ceiling. Allow 50–100 mm extra for the top, and the same for the bottom.

In most modern homes, you would normally get four drops per roll, depending on the pattern repeat. For properties with higher ceilings, or where a drop pattern is used, you may only get three drops per roll. Drops per roll need to be taken into account when you are measuring up, so cut a full roll, getting as many lengths as you can.

Activity: Measuring up

Measure the perimeter of your classroom or other small area, and then divide the measurement by the width of the paper (52 cm). This will tell you how many drops are required. Now work out how many rolls you need to decorate the room.

PLTS

Independent enquirer consider how different approaches could change the risks and hazards

Functional skills

English debate different approaches with others

Assessment activity 15.2 **P5** **P6** **M2**

BTEC

You have been hanging paper for a while, so your boss asks you to take special responsibility for health and safety issues applicable to the paperhanging team. She asks you to jot down some details of the different PPE and safe practices that are involved, as notes for a presentation to new staff and the rest of the team. Think about the hazards and risks associated with preparing walls and hanging wallpaper, then complete the following tasks.

1 List all of the PPE required and the safe working practices that should be followed when preparing walls and hanging vinyl wallpaper. **P5**

2 Explain why you have chosen specific items of PPE and justify the safe working practices that should be followed when completing this task. **P6** **M2**

Grading tips

1 To achieve **P5** you must identify all of the required PPE and safe working practices for the given task, thinking as broadly as you can.

2 To achieve **P6** you must explain why you have chosen the specific items of PPE, in terms of what the hazards could be.

3 To achieve **M1** you must justify the safe working practices identified, again looking at specific hazards you may come across.

Pasting

Apply paste down the centre of the paper and then out to the farthest edge in a criss-cross motion, making sure that you do not paste from the table edge inwards: this will result in paste getting onto the edge of the table and consequently onto the face of the paper. Allow the paper to soak for the time indicated on the manufacturer's label.

Papering

Place the paper on the wall, running it to the plumb line. Use a hanging brush to smooth the paper down, first down the centre and then out to the edges. You can use the brush to make any minor adjustments to keep the paper running to the plumb line.

Trimming

To trim paper, use a trimming knife and straight edge. When you move the straight edge across, always make sure that the knife blade stays in position. Remember to keep the blade sharp by snapping it off regularly. Once you have trimmed each length of wallpaper, it is important that you remove any paste from the adjacent surfaces, such as skirting boards, coving and door or window frames, using a sponge and clean water.

Papering internal angles

When papering into an internal angle, you need to measure from the edge of the previous length into the corner at the top, middle and bottom of the wall. It is important to measure at all three places, because the corner may not be plumb. You must then add 5 mm to the largest of these measurements, to allow the paper to go slightly round the corner and make sure that there is no gap in the corner. Once you have established the required width, you should then cut the paper to this width on the paste table before hanging.

When papering out of an internal angle, you must always mark a new plumb line. This makes sure that the pattern remains level with the previous wall, and helps to ensure that the pattern is still properly aligned when you get back to your starting point.

Once you have hung the two lengths in the corner, trim off any overlap around the corner to leave a neat finish.

Papering external angles

External angles, such as the outward edges of a chimney breast, are usually completed with one piece taken around the corner, ensuring that it is tight on the edge. If the paper creases as you take it round the corner, make a **splice joint** about 20 mm in from the external angle.

What is the best technique for applying paste to wallpaper?

Did you know?

If you put a small fold at the top and bottom of the length of paper after pasting it, it will stop paste from getting onto your hands, the woodwork and the ceiling.

Why is a plumb line called a plumb line?

Key term

Splice joint – when the paper is overlapped and a knife is run down, cutting through both layers to create a joint.

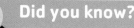

Papering fireplaces

Papering chimney breast walls is very popular at present as it creates a feature wall. Fireplaces can be difficult to paper due to the amount of complex cutting. To cut round a simple fireplace like the one in Fig. 15.6, there a number of steps to follow.

1 Paste the paper and position it on the wall above the fire surround, allowing the paper to hang over the surround.

2 Cut the paper in the direction of the arrows to the external points **A** and **B** shown in Fig. 15.6.

3 The cuts made will allow the paper to fall into place around the surround.

4 Smooth the paper down the wall toward point **C** and make an internal cut, following the direction of the arrow, to allow the paper to be folded into the corner at the skirting board.

5 Trim off the excess paper around the fire surround using shears or a knife and straight edge.

The same process applies to all fire surrounds; the external points and internal points need to be cut, to allow the paper to hang around the outline of the surround before trimming.

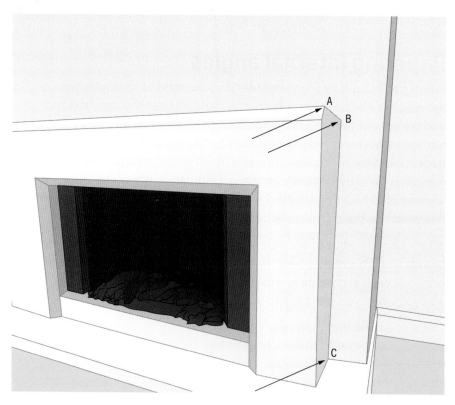

Fig. 15.6: The internal and external corners of a fire surround

Papering around windows and doors in reveals

Windows and doors set back into a wall are a different challenge, as a splice joint is required from the ceiling down to the corner of the **reveal**. This allows the paper to cover the top and the side of the reveal. You need to make cuts around the sill and a cut top and bottom of the reveal corner, to allow the paper to go round the corner to the window or door. Once the paper is smoothed out and creased into the edges, you can cut it as normal.

Papering around light switches and sockets

First make sure that the electricity is switched off. To paper around light switches and sockets, simply let the paper fall over the switch lightly and mark the four corners with a pencil. Ease off the paper and pierce the centre point of the switch with shears, next snip up to the corner points, taking care not to cut too far as the cuts will show when it is finished.

Key term

Reveal – the narrow piece of wall between the window or door frame and the outer surface of a wall.

Activity: Paperhanging process

Imagine you are employed by an interior design and decorating company. You have been asked to wallpaper a lounge (approximately 6m × 4.3m × 3.1m high) in a town house. The room has one large window, two doors set into wood-panelled reveals, and a picture rail. There is also a fireplace set into the chimney breast. The floor is hardwood, finished in clear varnish.

The walls have already been stripped, prepared and cross-lined, and the woodwork has already been painted. Your task is to hang the drop pattern vinyl wallpaper that the customer has selected.

- **What precautions will you need to take to protect the customer's property during the paperhanging process?**
- **What access equipment should you use?**
- **Where in the room will you hang the first length of wallpaper?**
- **What tools will you use to complete the task? State what each tool will be used for.**
- **How will you ensure the most economical use of the drop pattern wallpaper?**

PLTS

Reflective learner review your progress as you work

Effective participator plan your work step by step

Functional skills

Mathematics measure your overlaps accurately to stay within tolerances

:BTEC **Assessment activity 15.3** **P2** **P4** **P7** **P8** **P9** **P10** **M3** **D1**

Your final assessment for this unit will involve you completing a practical task. You will be required to prepare the walls in a room for paperhanging, cross-line them and hang a washable wallpaper to them.

1 Select the tools, access equipment and materials that you require to complete the given task. **P2** **P4**

2 Demonstrate that you can work safely, using the required hand tools and materials to complete the given task. **P8** **P9** **M3** **D1**

3 Demonstrate the safe use of low-level access equipment while completing the given task. **P10**

4 Throughout the completion of the given task, follow manufacturers' guidelines when preparing materials for use. **P7**

Grading tips

1 To achieve **P2** and **P4** you must take care to correctly select the required tools, materials and access equipment for the given task.

2 To achieve **P8** and **P9** you must complete the given task safely, using the correct hand tools and using materials in accordance with manufacturers' guidelines. To achieve **M3** you must produce all finished work with no air bubbles, blisters or wrinkles and no gaps or overlaps >2mm; to achieve **D1** your work must also have straight neat ends with no scissor marks, plumb deviation <2mm from vertical and any patterns accurately matched.

3 To achieve **P10** you must show that you can erect, use and dismantle low-level access equipment safely.

4 To achieve **P7** you must follow manufacturers' guidelines when mixing fillers and pastes.

Ben Edwards

Decorator

Ben is a qualified decorator working for a company specialising in high-quality domestic work. He has been interested in decorating since he completed a BTEC First Diploma in Construction while still at school. After leaving school, he completed an apprenticeship as a decorator with his current employer.

Ben particularly likes the paperhanging aspect of his work and has always sought opportunities to work on jobs where the customer has specified wallpaper. This means that he has gained a lot of valuable experience with a wide range of different wallpapers. Recently, he has been using some specialist wallpapers including Lincrusta™, flock and contract-width vinyl. Ben has really enjoyed using these materials and he is now sure he would like to become a specialist in this area.

Ben has done a bit of market research and discovered that there are not many people locally who can provide a service hanging some of the more expensive and specialist wallpapers. However, there seems to be an increasing demand for this type of work and he has decided to fill that gap in the market. To do this, he will be setting up his own business as a specialist decorator.

In order to set up his own business, Ben feels that he needs to gain a better understanding of business and things like finance, tax and taking on employees. He feels that a good starting point would be to contact his local college to ask about short courses for new businesses.

Think about it!

1 What courses are available at your local college that would help Ben to get started in business?

2 What do you think would be the potential advantages and disadvantages of Ben taking on an apprentice?

3 How could Ben promote his new business?

Just checking

1 What is meant by a porous surface?
2 What is the purpose of a plumb line and bob?
3 Name two tools used for cutting wallpapers.
4 When would the decorator not use starch paste?
5 What is the main advantage of using an Anaglypta™ or blown vinyl wallcovering?
6 Draw the symbols for a straight and a drop match paper.
7 What is the size of a standard roll of wallpaper?
8 What is the hazardous ingredient that most pastes contain?
9 State the main reason for cross-lining.
10 State one safety action you would perform after stripping old wallcoverings or using paste.

edexcel

Assignment tips

- Look around you when you're out and about and check out how many different types of wallpaper you can see in different places.

- Why not visit a stately home or two and see how wallpapers have changed over the years?

- Pop into your local trade centre and look through the wallpaper books. See how much some of the top-end wallpapers cost compared to wallpapers at your local DIY centre. Are there many that require special treatment when hanging?

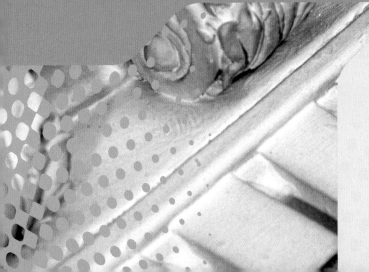

16 Performing decorating operations

Paint and wallpaper are not the only ways to make rooms nicer to live in. Coving and centrepieces can add a touch of elegance to a room. Likewise, textured finishes can completely transform a room. Texturing materials are versatile and can be used on walls and ceilings to produce a wide range of patterns and finishes that are hardwearing and easy to redecorate when a change of colour is required.

Skills and knowledge about installing coving, centrepieces and textured finishes will be attractive to a prospective employer. They will also form a strong foundation, from which you can further develop as you learn more about these forms of decoration.

Of course, dealing with coving, texturing and centrepieces are not the only skills a painter and decorator needs.

You may study this unit as a stand-alone topic, but you can also combine it with others such as Unit 14 Exploring painting and decorating and Unit 15 Performing paperhanging operations.

Learning outcomes

After completing this unit, you should:

- know the hand tools, materials and access equipment used by decorators
- understand safe working practices for the application of textured finishes and the installation of coving and ceiling centrepieces
- be able to apply safe working practices when performing coving, texturing and ceiling centrepiece installation tasks.

Assessment and grading criteria

This table shows you what you must do in order to achieve a pass, merit or distinction grade, and where you can find activities in this book to help you.

To achieve a **pass** grade the evidence must show that you are able to:	To achieve a **merit** grade the evidence must show that, in addition to the pass criteria, you are able to:	To achieve a **distinction** grade the evidence must show that, in addition to the pass and merit criteria, you are able to:
P1 identify the hand tools and access equipment used to perform decorating tasks **See Assessment activity 16.1, page 208**	**M1** justify the safe use of tools, materials and access equipment to minimise health, safety and welfare risks **See Assessment activity 16.1, page 208 and Assessment activity 16.2, page 213**	
P2 select hand tools and access equipment to perform specified decorating tasks **See Assessment activity 16.2, page 213**		
P3 identify materials used in decorating tasks **See Assessment activity 16.2, page 213**		
P4 select materials used in specified decorating tasks **See Assessment activity 16.2, page 213**		
P5 identify the PPE and safe working practices used to perform specified tasks **See Assessment activity 16.1, page 208**	**M2** justify the use of appropriate PPE and safe working practices to minimise health, safety and welfare risks **See Assessment activity 16.1, page 208**	
P6 explain the selection of PPE and safe working practices used to perform specified tasks **See Assessment activity 16.1, page 208**		
P7 follow manufacturers' guidelines when preparing materials for use **See Assessment activity 16.3, page 214**	**M3** produce finished work with neat and consistent textured finishes and covings and ceiling centrepieces accurately placed to ±3 mm **See Assessment activity 16.3, page 214**	**D1** produce finished work with neat and consistent textured finishes, covings and centrepieces accurately placed to ±3 mm, a high standard textured finish with no misses or other imperfections visible from 1 m, and covings and ceiling centrepieces accurately placed to ±1 mm **See Assessment activity 16.3, page 214**
P8 perform specified tasks using hand tools **See Assessment activity 16.3, page 214**		
P9 demonstrate the safe use of materials when performing coving, texturing and ceiling centrepiece installation tasks **See Assessment activity 16.3, page 214**		
P10 demonstrate the safe use of low level access equipment when performing coving, texturing and ceiling centrepiece installation tasks **See Assessment activity 16.3, page 214**		

How you will be assessed

This unit will be assessed by an internal assignment that will be designed and marked by the staff at your centre. Your assessment could be in the form of:

- presentations
- case studies
- practical tasks
- written assignments.

Tom, 16 years old

Before starting this unit I didn't realise that installing coving and applying textures was the job of a decorator. I knew decorators applied paints and wallpaper, but I'd never really thought about the other features used to decorate places. This unit has been a great way to increase my knowledge of decorating and the skills involved.

I learned about tools I'd never heard of before and put into practice what I learned. Knowing how to produce a range of effects and finishes, and install coving and centrepieces is great because they can bring a whole new look to a room, even when they're really simple. I enjoyed the practical aspect of this unit most of all. For an assessment I put up coving and a centrepiece and finished them off with emulsion. It was great to see the final result and achieve a quality finish.

This unit has made me consider a career as a decorator because it's shown me how much variety there is in painting and decorating – you can use such a wide range of techniques and materials just to complete one room.

Over to you

- Which areas of this unit might you find challenging?
- Which particular section are you looking forward to most?
- What can you do to prepare yourself for the unit assessment?

1. Hand tools, materials and equipment

Build up

Changing rooms

Did you know that, until the late 1960s, texturing materials contained asbestos?

Why do you think that asbestos is no longer used in texturing materials and other building products? Think about other decorating materials that have gone in and out of fashion. Why is this so?

1.1 Hand tools

Most of the tools used for texturing and coving are unique to each of the applications. For texturing, coving and centrepieces to be successful, your preparation has to be correct.

Tools for texturing

The set of tools for texturing are different to those used for putting up coving and centrepieces.

Bumper

This is used for mixing texturing material, by plunging it up and down while turning at the same time. The tool has a broom-type handle, and the end is a circular disk with holes to let the material pass through. This helps to break down lumps in the material to ensure a smooth mix.

Bark roller

This has a normal roller frame, but the head is made from plastic or rubber. You can use it to produce a bark effect in texturing material, by rolling it down the wall vertically.

Lacer

A lacer consists of two differently-sized flexible triangular blades riveted together. The smaller one is used for intricate areas. It is passed over a textured surface to remove sharp points once the texturing has been done but before it has set.

Laying-in brush

This brush has a thick-set head to ensure thick application, and is used to apply the texturing material to walls or ceilings.

Laying-in brush

Rubber stippler

A stippler is made of rubber strands set into a plastic or wooden block, with a handle on the top. This is used to produce standard stipple and swirl effects.

Texturing combs

A texturing comb is a flexible plastic blade with grooves or curves cut into the edge. It has a shorter blade next to the main blade to ensure even pressure across the blade when you use it. Texturing combs are used to produce various patterns once the texturing material has been applied.

Caulking tool

This consists of a wide plastic blade set in a wooden handle, and is used for applying filling materials to plasterboard joints before applying any texturing materials.

Textured rollers

These have rubber or foam roller sleeves with sculptured patterns on them, and are used to create a variety of patterns, depending on the design. The texture is applied by brush and the roller is rolled over the surface, creating the effect.

Tools for coving and centrepieces

To install coving and centrepieces you must use tools which allow you to cut and measure accurately and apply adhesive.

Coving mitre

This is made from metal or plastic, the metal being much sturdier than plastic, and is used to get the correct angle for Gyproc coving. It works for 90° internal and external corners and also for a mitred joint and is placed over the top of the coving as a guide for the saw.

Hard point saw

This is a metal blade with serrated teeth, used for cutting mitres and joints in coving.

Rubber stippler

Texturing combs

Coving mitre

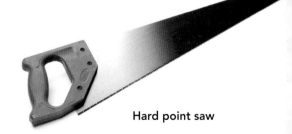

Hard point saw

Tape measure

Filling knife

Trimming knife

Tape measure

Tape measures are produced in various sizes, usually between three and ten metres, and most contain metric and imperial measurements. You use them to produce accurate measurements before cutting the coving.

Filling knife

A filling knife has a flexible blade and tang made from one piece, with a hardwood handle. You use it to create a sharp top and bottom edge on the coving, by running the blade along the wall and ceiling, creating the edge.

Trimming knife

This has an extremely sharp blade that can be snapped off up to ten times using the slot at the opposite end. It is used to trim any excess from the length and also to trim the rough edge left by the saw.

1.2 Materials

Texturing materials

Texturing materials are usually powder-based, and are available in 10 kg and 25 kg bags. They are available as a cold- and a hot-water mix; however, all-temperature mixes are also available. Ready-mixed is also available but this is more expensive and does not give the control of consistency that powders offer. To mix it, you should use a clean bucket or large container, adding the powder to the water and mixing constantly until the desired consistency is achieved.

Today texturing material is used mainly on ceilings, although it was popular on walls until the late 1980s. It gives a textured look to walls and ceilings and, once dry, is painted over to leave a hard-wearing surface. It is also used to cover surfaces that are in poor condition. Texturing material is used in new builds on ceilings, applied directly onto the plasterboards because it is less expensive than having them skimmed with a layer of plaster.

Activity: Working out quantities

You are required to texture the ceiling in your classroom or workshop.

- **Calculate the area in square metres.**
- **How much texturing material would you require if 1 kg covers 2 m²?**
- **Which size bags would you use to make the most economical use of the material if 10 kg costs £10.50 and the 25 kg bag costs £18.50?**

Plasterboard coving

Plasterboard coving or Gyproc coving is gypsum plaster that is covered with a layer of paper during manufacture, and has the same composition as plasterboard. It is used to enhance any room by providing a decorative finish where the ceiling and walls meet. It can also be used to hide bad cracks between the ceiling and walls, and to hide wiring if, for example, spotlights are to be installed in a lounge area.

Plasterboard coving has a C-shaped profile (any other profile is known as cornice). It can be purchased in 2000 mm lengths, but people in the trade use 3000 mm, 3600 mm and 4200 mm lengths to minimise joints and increase installation speed. Two profile widths are also available: 127 mm, which is the most common type used and comes in the lengths mentioned, and 100 mm, which is only available in 2000 mm and 3000 mm lengths. The smaller width can suit small rooms or rooms with low ceilings.

Fibrous plaster cornice

Fibrous plaster cornice is available in a range of decorative styles, from plain to highly decorative. It is made from plaster of Paris with cotton scrim and thin timber battens set into the plaster for reinforcement. Length and depth is governed by the size of the mould used in production. In stately homes it is common to see the pattern highlighted with gold leaf.

Polystyrene coving

Polystyrene coving is a lightweight alternative to the plaster types. It is usually covered with a paper to give it the same look as the plaster type. As it is lightweight, it is easy to put up but it can be easily damaged and dents readily. It is only available in limited designs and smaller sizes.

Centrepieces

Centrepieces can be made of either plaster of Paris or foam. The main difference is that the foam variety is lightweight and can be installed using only adhesive; the plaster type is much heavier and usually requires the use of screws and adhesive to secure it to the ceiling. Both types can add decorative elegance to a room; usually the design will match or complement the cornice. Centrepieces are available in various diameters, starting at around 200 mm. You must take care to choose the right size for the room: too big, and it could look overpowering; too small, and it could look lost on a large ceiling. Once installed and painted, the two types of centrepiece look identical.

1.3 Personal protective equipment

When performing texturing and coving operations, you must wear personal protective equipment (PPE). Essential items of PPE for this type of work include, safety footwear, overalls, gloves, safety glasses, when mixing and applying overhead, and a dust mask, when mixing powder-based materials and cutting gypsum-based materials.

> **How can coving help to finish off the look of a room?**

> **In what kind of homes would you expect to see ornate cornices?**

Did you know?

Plaster of Paris is calcium sulphate (gypsum) that is heated to a very high temperature and is a class A hemi-hydrate. It is also classed as a chemically active surface.

Cross reference

Look at Unit 2 for more information about PPE.

1.4 Access equipment

'Working at height' means working at any height that is off the ground, even if that is only 100 mm. Due to the nature of decorating operations, most work is done from access equipment, including step ladders, hop-ups and trestles with staging.

If you are texturing a ceiling and putting up coving, you need to create a run across the width of the room with trestles and a lightweight staging, to make sure that you can work safely and efficiently. It would be extremely difficult to put up a length of coving of three metres or more, working alone on a pair of steps, without dropping it. Similarly, when texturing a ceiling, if you just use a pair of steps, you would be constantly up and down, which would resulting in your work having dry edges.

When installing centrepieces, a pair of steps is usually enough. Stepladders must be fully open when in use and must be of the correct size for the job, ensuring your knees are not above the top tread.

Fig 16.1: Trestle working platform

BTEC Assessment activity 16.1

You have gone to look at a room where you will be installing some coving and a centrepiece, as well as applying a textured swirl finish to ceiling. You have been asked to assess the room and the job, looking at any potentially hazardous materials and other risks, and to report on the PPE, tools and access equipment needed for the job.

1   State what PPE you will need when installing the coving and centrepieces and why. **P5 P6 M2**

2   List the tools and access equipment you will need when applying a textured finish to a ceiling area, and explain why you need each one. **P1 M1**

Grading tips

1   To achieve **P5** you should identify all the required PPE items. To achieve **P6** you should explain why you have selected each item of PPE.

2   To achieve **P1** you should identify all of the hand tools and access equipment required to complete the task. To achieve **M1** and **M2** you should state why you have selected each hand tool and item of access equipment, relating them to the risks and hazards involved.

2. Safe working practices for the application of textured finishes, and the installation of coving and ceiling centrepieces

2.1 Textured finishes

The application of texturing material should be a safe task, providing you take a few precautions. As texturing material is chemically active, you should apply barrier cream to prevent drying of the skin.

You should wear a dust mask when you are mixing, as you will usually be working with a powder-based material. If possible, try to mix the texturing compound outside: this stops the dust getting inside and means there is less chance of the customer's property being damaged. Wherever you do your mixing, use a large board or polythene sheet under the container to protect the floor.

Texturing material tends to be sloppy; when applying it to ceilings, it is advisable to wear safety glasses or goggles to protect your eyes. When texturing large quantities of material are required and the containers can be heavy, so think about splitting the amount into two; this will reduce the risk of back injuries and ensure better balance and better carrying posture.

2.2 Fixing covings

Coving can make a big difference to a room. It provides a neat ceiling line and gives a good edge for cutting in with paint, or trimming up to when wallpapering.

Measuring and cutting

When you are performing coving operations, accurate measurement is essential to ensure a high-quality result. If you are using 127 mm coving, as shown in the photos overleaf, you need to measure 85 mm down from the ceiling and mark a horizontal line on the wall for the coving to follow. This measurement allows for the thickness of the adhesive you will apply to the back of the coving. When using coving of different depths, you will need to alter this measurement.

Next you need to fix the coving. Take the measurement of the first wall minus 2–3 mm. Decide whether you require internal or external cuts, then place the coving mitre on the coving. Make sure your saw is lined up to cut down to the measurement mark. Remember that the saw is cutting at an angle, so take this into account when positioning it. Place the saw so it is resting against the coving mitre and keep it there as you cut through the coving. Once you have cut it, it is good practice to try the piece in place before applying the adhesive.

Did you know?

When coving has been cut using a saw, there is a rough edge, which can make it difficult to create a nice, clean joint. Smoothing the edge with a trimming knife or abrasive paper will result in a cleaner joint that is easier to fill and looks sharper.

When the coving is in place, how can you neaten up any excess adhesive?

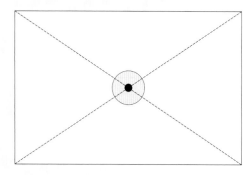

Fig 16.2: Finding the centre of a ceiling

Mixing and applying adhesive

Mix up the coving adhesive to a medium stiff paste and apply it to the back edges of the coving with a broad knife. Remember that only the top and bottom edges adhere to the wall and ceiling; the middle section does not require adhesive.

Once you have applied the adhesive, it is important that you take care when climbing onto your working platform. Place the coving in position and align it with the marks you made earlier. Take care to ensure a good fit with adjoining pieces. Once it is in the correct place, press the coving firmly to squeeze out any excess adhesive and ensure a strong bond between the two surfaces.

Always make sure that you follow the horizontal line to ensure the last piece aligns with the first length. On long runs, it is a good idea to use a few nails, spaced along the setting-out line, to support the coving and reduce sagging in the middle. This is particularly helpful if you are working alone.

Once the coving is in place, any excess adhesive can be removed from the ceiling and wall using a filling knife. This creates a nice, sharp angle between the coving and the wall or ceiling. Any gaps between the coving and the wall or ceiling can be filled with adhesive using a putty knife or filling knife. Once this has been done, a wet brush can be used to clean the surfaces. The brush should be held at 90° to the wall or ceiling thus helping to keep the joint flat and avoid any sinking of the adhesive.

When using a coving mitre to cut internal angles, the off-cut could be used for an external angle if it is long enough.

Fibrous plaster coving and cornice

Unlike plain Gyproc coving, most fibrous plaster covings and cornices have ornate patterns that need to be matched when installing them. This type of cornice is much heavier and requires more care when handling, as it is easily damaged. Also, because of the extra weight, it normally requires the use of screws and adhesive to hold it in position. You should also take care when lifting and carrying it.

2.3 Fixing centrepieces

When installing centrepieces, it is important to find the exact centre of the ceiling. You can do this by snapping two chalk lines diagonally from corner to corner, as shown in Fig 16. 2. The centre of the ceiling is the point at which they cross over.

Sometimes the centre of the ceiling is indicated by the location of the light fitting. In larger rooms, there may be two light fittings in the ceiling; in such cases, you would usually install two centrepieces.

Foam centrepieces are lightweight and relatively easy to install, requiring only a coving-type adhesive; plaster centrepieces are much heavier and require the use of screws into the ceiling joists in addition to the adhesive. For both types, you will need the services of an electrician if a light fitting needs to be removed. Once you have positioned and secured the centrepiece correctly, remove any excess adhesive and fill any gaps using the same processes as when installing coving (see page 210).

2.4 Health, safety and welfare

When installing coving and centrepieces and applying textured finishes it is vitally important to ensure that your work area is kept as clean and tidy as possible. Tools, materials and equipment that are left lying around in a workspace can create trip hazards and spilled materials such as adhesive, water or texturing materials can also cause slip hazards. A clean and tidy work area also enables you to move yourself and your access equipment around more easily and more safely. This also provides for more efficient working practices.

When performing these tasks it is important to carry out a risk assessment before starting work. A good risk assessment will identify hazards, risks, the likelihood of injury or illness occurring, those at risk and control measures to reduce the risk.

Hazards associated with fixing coving and centrepieces

The adhesive used to fix coving and centrepieces is chemically active and you should take the same precautions as with texturing materials when using it.

As centrepieces can be heavy, it is essential that you use the correct manual handling techniques. Use kinetic lifting techniques when lifting, and always split the load if it is too heavy.

When working with coving, take care when using a saw to cut the mitres and joints. Use the coving mitre and a good platform for cutting, to help reduce the risk of injury.

Hazards associated with access equipment

When setting up access equipment, especially for ceilings, make sure it is high enough that you are not stretching; this will prevent you getting back problems and frozen shoulder.

Did you know?

When screwing fibrous plaster cornices to the wall and ceiling, make sure you use non-ferrous screws as they will not rust and 'bleed' through the paint system.

Key term

Bleeding – when rusting metal creates a reddish brown stain on a surface.

Activity: Kinetic lifting

In pairs, discuss what is meant by the term 'kinetic lifting', then list as many good lifting practices as you can.

COSHH

The Control of Substances Hazardous to Health (COSHH) regulations are designed to protect the employee. They place the onus on the employer to:

- know the materials being used
- assess the damage to health that the substances can cause (risk assessment)
- decide how to prevent or control them
- provide information and training to employees.

The employee must use the control measures that the employer has put in place when using materials that are hazardous to health.

3. Safe working practices when performing coving, texturing and ceiling centrepiece installation tasks

3.1 Preparing bare and previously painted surfaces

Surface preparation is essential for a successful job. Table 16.1 below gives you a guide through the preparation needed for the type of work covered in this unit.

Table 16.1: Preparing surfaces

Surface	Condition	Preparation required
Plaster	New and unpainted	A coat of thinned-down PVA adhesive is essential to reduce the porosity of the surface.
	Painted with water-based matt in sound condition	As water-based matt emulsion is porous, a thinned coat of PVA adhesive is required to reduce the porosity and aid the bond of the materials to the surface. When installing coving and centrepieces, you must score the surface in a criss-cross fashion using a knife or the edge of a scraper; this will help the adhesive to hold.
	Painted with a solvent-based eggshell or gloss	This is a non-porous surface, but a key is needed to bond to this type of surface. This can be achieved by washing it with sugar soap and rubbing it down using aluminium oxide grade 80. When installing coving and centrepieces, you must score the surface in a criss-cross fashion using a knife or the edge of a scraper; this will help the adhesive to hold.
Plasterboard	New and unpainted	Joints must be taped with self-adhesive fibreglass jointing tape to prevent cracking.

3.2 Applying textured finishes

When mixing texturing material it is essential to mix a sufficient amount to cover the entire surface with one mix, thus ensuring the same consistency throughout the area you are texturing. If you have to mix more material part way through a wall or ceiling, the edge will be drying and the consistency of the new material could be slightly different, resulting in a visible difference where the mix was changed.

There are many texturing designs. Four of the most popular include swirl, broken leather, bark and stipple.

How is this texture created?

Swirl pattern

A swirl pattern is produced by laying on the texturing mix and then rotating the rubber stippler through 180° and lifting off the surface in one smooth movement. This leaves a stipple pattern in the centre of the swirl. A sponge wrapped in a plastic bag can also be used to give a similar effect.

Broken leather effect

You can achieve this effect by applying the texturing material reasonably thickly, and then wrapping the brush in a plastic bag. Pass the brush over the surface in random directions to create the desired effect.

BTEC Assessment activity 16.2 (P2) (P3) (P4) (M1)

You have been given the task of applying a broken leather patterned textured finish to a large ceiling area in a modern domestic property.

1 From the full range of texturing tools available, select the ones you will need to produce the stated finish. (P2)

2 Select the appropriate access equipment for the task and state your reasons for selecting each item. (P2) (M1)

3 State what materials you will need to complete the given task. (P3) (P4)

Grading tips

1 To achieve (P2) you must select all of the required tools and access equipment for the specified task, thinking about how to achieve that particular effect.

2 To achieve (P2) you must select all of the required access equipment for the specified task, thinking about working on a ceiling. To achieve (M1) you must state why you need the tools, materials and access equipment that you have selected.

3 To achieve (P3) you must identify what materials are needed to complete the task, and to achieve (P4) you must select these correctly.

PLTS

Reflective learner ask your tutor for feedback on how you are doing

Functional skills

ICT use the internet to find images of broken leather effect in situ

Bark effect

This pattern is produced by applying the texturing compound to the wall using the laying-in brush. You must do this as evenly as possible to keep the pattern the same throughout the job. Once the material has been applied, roll the bark roller, which is completely smooth, down the wall vertically to create the texture, and repeat in rows until the wall is complete.

Stipple

A regular stipple is the easiest pattern to produce, and is the pattern that is used the most often, especially on ceilings. Apply the texturing material to the ceiling or walls as evenly as possible to create a uniform pattern, then dab it all over using the rubber stippler.

When applying textured finishes, it is essential to run a wet brush around all of the edges after the pattern has been produced. This is because the pattern cannot normally be produced in the angles and the wet brush will ensure a uniform finish at the edges of the surface.

PLTS

Creative thinker think about the style of coving that best suits the space

Team worker work with others to install the centrepiece

Functional skills

Mathematics measure carefully in order to position the coving properly

BTEC Assessment activity 16.3 (P7) (P8) (P9) (P10) (M3) (D1)

Your final assessment activity for this unit involves applying textured finishes, installing coving and fixing centrepieces. You will be given a work area that consists of a ceiling and walls that have at least one internal and one external angle.

1 Install plasterboard coving to the work area, ensuring that you include one internal and one external angle. (P8) (M3) (D1)

2 Produce a bark-patterned finish to one wall in your given work area. (P8) (M3) (D1)

3 Install a ceiling centrepiece in your given work area. (P8) (M3), (D1)

4 In order to complete the above tasks you will need to select and safely use appropriate access equipment. (P10)

5 You must follow all manufacturers' instructions whilst preparing materials for use and throughout their application. (P7) (P9)

Grading tips

1 To achieve (P8) and (M3) you must complete the given practical tasks, and your finished work must be to an acceptable standard; to achieve (D1) your finished work must be to a good, neat standard.

2 To achieve (P10) you must show that you can use the access equipment safely.

3 To achieve (P7) you must follow manufacturers' instructions while you are preparing the materials for use. To achieve (P9) you must demonstrate that you can use the materials safely.

Joanna Flowers
Decorator

Joanna is an experienced decorator who works for a national house builder painting new homes. The work consists mainly of applying a stippled texture to the ceilings and magnolia emulsion to the walls throughout each house, and a plain Gyproc coving to the lounge, dining room and master bedroom.

Joanna served her apprenticeship with the company and stayed with them for the five years since completing it. While she enjoyed some of the work, particularly the coving and texturing, she felt increasingly that she needed a new challenge.

In her spare time, Joanna likes to visit stately homes and has a strong interest in period decoration. She particularly likes to see ornate cornices and ceiling centrepieces. It is this interest that spurred her on to look for a new employment opportunity.

Twelve months later, Joanna is now working for a small company based in the north of England that works nationwide on restoration and refurbishment projects. Her main job is installing fibrous plaster decorative cornices and ceiling centrepieces. The existing cornices can often be repaired in situ, and this work also falls to Joanna.

Joanna is considered to be very good, but she would like to undertake some formal training so that she can gain accreditation in this new line of work. One possible route for Joanna to follow is to complete an NVQ in Plastering, which has a number of different specialist routes, including Fibrous Plastering.

One distinct advantage of taking the NVQ route is that she could do much of her learning and assessment in the workplace.

Think about it!

1 What would be the long-term benefits to Joanna of gaining additional qualifications?

2 How would Joanna's site experience help her in her further studies?

3 What other training or qualification routes could Joanna take?

Just checking

1 Name three tools used for applying textured finishes and describe their uses.
2 How should you treat new plaster before applying a textured finish?
3 Name two different patterns that can be produced using textured finishes.
4 How should the edges of a textured surface be finished?
5 Why is a horizontal line required when installing coving?
6 What tool is used to create a sharp edge on the ceiling and wall when installing coving?
7 For what purpose would nails be required when installing coving?
8 What, in addition to adhesive, do you need to fix fibrous plaster coving?
9 How can you accurately identify the centre of a ceiling?
10 Whose services would be required when installing a centrepiece around a light fitting?

edexcel :::

Assignment tips

- Visit your local library or town hall. You will usually find lots of good examples of decorative cornices and ceiling centrepieces. See how many different patterns you can find on the cornices.

- Apart from the range of patterns available, think about different reasons why textured finishes might be used in a building.

- Compare old and modern buildings and look at the differences in how they are decorated. Make notes on any queries you have to ask your tutor about.

Glossary

A

Abraded – rubbed to smooth the surface and provide a key

Added value – a value that has not cost you anything directly but which has provided a second value to a project

Aesthetics – a pleasing or beautiful appearance

Angle of repose – the natural angle that a substance settles itself at if you pour it onto a surface

Apprenticeship – a programme of learning and qualifications, completed in the workplace and college or training centre, that gives young people the skills, knowledge and competence they need to progress in their chosen career or industry

B

BA – breathing apparatus

Biodiversity – the range of plants and wildlife within a region

Bleeding – when rusting metal creates a red stain on a surface

Bond – a technique of laying bricks on top of one another in a pattern, so that one brick always overlaps another

Burning off – the removal of old paint films by the application of heat from a blowtorch or hot air stripper, making the paint blister so that it can be scraped off

Butt joint – when the edges of the paper are touching but not overlapping

C

Carbon neutral – neither adding carbon to the atmosphere nor taking it away

Carbon quota – the amount of CO_2 (measured in tonnes) that a country can emit from industrial processes under international agreements

Competent – having the qualifications, training, knowledge and experience to be able to do something

Conurbation – an extensive urban area formed by growing cities

Corner – a 90° change of direction of a wall

COSHH – Control of Substances Hazardous to Health

Course – a row of bricks or blocks

Cross-line – hang lining paper horizontally to give an even, more professional finish when the wallpaper is hung over it

Cross-section – what the inside would look like if you cut through it

D

Datum – a level given which all other levels relate to

De-nib – lightly rub down between coats of paint to ensure a smooth surface

Delamination – when the two layers of a laminated paper separate, usually occurs on the edges

Drop match – when the pattern does not match horizontally edge to edge

E

Egress – going out, exiting

Embossed – with a raised pattern

Encapsulate – cover or enclose completely without disturbing

Enforced – something that must be obeyed or prosecution could take place

F

Facilities – any activity required for the running and operation of a building

Ferrule – a metal band that secures the bristles to the handle

Finite – things that cannot be replaced once used

Fossil – a natural carbon-based material

Free or random match – a design that does not require matching as there is no set repeat

G

GDP – gross domestic product, the amount of 'income' a nation has from what it produces and sells

Gypsum – calcium sulphate

H

HASAWA – Health and Safety at Work etc. Act (1974)

Header – a brick laid so that its short side is parallel to the face of the wall

HSE – The Health and Safety Executive

I

Impermeable – will not let water absorb into it or pass through it

Ingestion – taking into the body by eating or swallowing

J

Justify – give a good reason for

K

Key – a scratched pattern in the surface of the block that makes the plaster coat adhere to the concrete block

L

LBC – London Brick Company

Lift – the amount of brickwork that can be built before scaffolding is required or needs to be moved because you cannot reach

M

Monitoring – the continual measurement of a system

Mortar – holds the individual bricks and blocks together

Musculoskeletal – to do with the human frame and muscles that functions to give movement

O

Opacity – the quality of lacking transparency, so that you cannot see through it

P

Pile length – the length of the material used for the roller sleeve. Long-pile rollers are best for heavily textured or rough surfaces; short-pile rollers give the best results on smooth surfaces

Plasticisers – modern chemicals that allow grains of sand to slide past each other easily

Porosity – how porous or absorbent something is

PPE – personal protective equipment

Professional – a person whose occupation requires specialist learning

Psychosocial – to do with the way in which you interact subconsciously with your environment: for example, when you are tired from working without a break, then you slip and fall, causing personal injury

R

Rating – a measure of how severe a hazard has the potential to cause harm

Regeneration – redevelopment of land or the upgrading of older, rundown areas

Relevance – importance for and connection to something

Reveal – the narrow piece of wall between the window or door frame and the outer surface of a wall

Roller sleeve – the soft part of a paint roller that holds the paint during use, usually made of lambswool, mohair or a man-made material

S

Second fixings – the secondary items, such as mouldings, that are mainly added after plastering has finished

Self-neutralising – for a paint remover, when the liquid neutralises itself, so the surface you use it on does not need further washing afterwards

Severity – the extent to which something is bad, serious or unpleasant

Simplex – made of a single layer of paper, such as standard or washable paper

Sizing – applying a special coating material to a surface to make it less absorbent

Smog – fog that has become mixed and polluted with smoke

Sound surface – a surface that has previously been painted and is in generally good condition

Splice joint – when the paper is overlapped and a knife is run down, cutting through both layers to create a joint

Springing – coming unstuck and sticking out

Straight match – when the pattern matches on a horizontal line

Stretcher – a brick laid so its long side runs parallel to the face of the wall

SUDS – Sustainable Drainage Systems, which aim to decrease and slow down surface run-off, or divert it for other useful purposes

Surveying – measuring an area to establish its size and shape

Swale – a hollow or marshy depression between ridges

T

'T' junctions – a wall that meets the outer wall at a junction, forming the shape of the letter T. This occurs at the junction of the internal and outside walls

Tang – the section of metal inside the handle of a scraper or knife

Tetanus – a rare but often fatal disease that affects the central nervous system, caused when bacteria enter the body through a wound or cut

U

UCAT – Union for Construction and Allied Trades

Unsound surface – a surface on which the paint film has broken down severely over most of its area

W

WAH – Working at Height

Workability – the quality of being flexible and easy to move about

Index

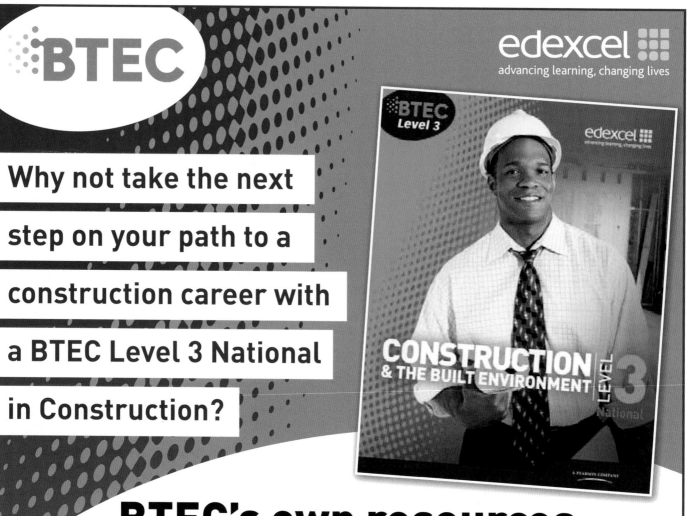